T0212239

Trading Agents

Synthesis Lectures on Artificial Intelligence and Machine Learning

Editor
Ronald J. Brachman, *Yahoo Research*
William W. Cohen, *Carnegie Mellon University*
Thomas Dietterich, *Oregon State University*

Action Programming Languages
Michael Thielscher
2008

Representation Discovery using Harmonic Analysis
Sridhar Mahadevan
2008

Essentials of Game Theory: A Concise Multidisciplinary Introduction
Kevin Leyton-Brown and Yoav Shoham
2008

A Concise Introduction to Multiagent Systems and Distributed Artificial Intelligence
Nikos Vlassis
2007

Intelligent Autonomous Robotics: A Robot Soccer Case Study
Peter Stone
2007

© Springer Nature Switzerland AG 2022

Reprint of original edition © Morgan & Claypool 2011

All rights reserved. No part of this publication may be reproduced, stored in a retrieval system, or transmitted in any form or by any means—electronic, mechanical, photocopy, recording, or any other except for brief quotations in printed reviews, without the prior permission of the publisher.

Trading Agents

Michael P. Wellman

ISBN: 978-3-031-00426-1   paperback
ISBN: 978-3-031-01554-0   ebook

DOI 10.1007/978-3-031-01554-0

A Publication in the Springer series
SYNTHESIS LECTURES ON ARTIFICIAL INTELLIGENCE AND MACHINE LEARNING

Lecture #12
Series Editor: Ronald J. Brachman, *Yahoo Research*
                William W. Cohen, *Carnegie Mellon University*
                Thomas Dietterich, *Oregon State University*
Series ISSN
Synthesis Lectures on Artificial Intelligence and Machine Learning
Print 1939-4608   Electronic 1939-4616

Trading Agents

Michael P. Wellman
University of Michigan

SYNTHESIS LECTURES ON ARTIFICIAL INTELLIGENCE AND MACHINE LEARNING #12

ABSTRACT

Automated trading in electronic markets is one of the most common and consequential applications of autonomous software agents. Design of effective trading strategies requires thorough understanding of how market mechanisms operate, and appreciation of strategic issues that commonly manifest in trading scenarios. Drawing on research in auction theory and artificial intelligence, this book presents core principles of strategic reasoning that apply to market situations. The author illustrates trading strategy choices through examples of concrete market environments, such as eBay, as well as abstract market models defined by configurations of auctions and traders. Techniques for addressing these choices constitute essential building blocks for the design of trading strategies for rich market applications. The lecture assumes no prior background in game theory or auction theory, or artificial Intelligence.

KEYWORDS

trading agent, bidding agent, algorithmic trading, auction, computational game theory, eBay, multiagent systems

Contents

Preface

Delegating decisions to automated software agents is both compelling and fraught with risk. Computers can process massive volumes of information, respond in microseconds to events, and reliably follow instructions despite their tedium. But automation requires that we provide in advance the right instructions for the situations that will arise in the complex environments in which we set the agents loose. By including in the agent adaptive components that are shaped by the agent's own experience, we may improve its flexibility and potential performance, but at the same time increase the difficulty of predicting its behavior.

When the decisions are about trading in markets, the stakes are significant on both sides of the automation dilemma. The vast amount of data and complexity of calculations bearing on the trading decision—valuing goods and services, predicting future market prices—necessitate algorithmic assistance. Rapidity of response required by dynamics of the market may preclude direct human involvement. On the other hand, trading decisions inherently produce economic consequences. Delegating authority to trading agents means trusting to software the disposition of real resources—money, commitments to provide goods or services, or whatever is being exchanged in the market at hand.

This lecture is designed for those who would design trading agents. Whereas it cannot address the general problem of automating complex decisions, the lecture instead focuses on issues specific to market environments. By presenting key principles drawn from research in economics and artificial intelligence, it aims to equip the designer with some basic tools for strategic reasoning about market interactions. By "tools" I refer to patterns of thinking that support strategic analysis of well-specified market scenarios. We derive these patterns through study of canonical market forms, and identification of pivotal strategic questions that come up in market situations. Although the markets and situations we explore are inevitably idealized simplifications, the concepts and insights we take as lessons from these studies form the building blocks for more comprehensive agent design.

Much of our study of canonical market models lies within the realm of *auction theory*. There are many existing surveys and even textbooks covering this ground, and numerous references to same are provided within. The vantage taken in this lecture differs from standard treatments in two broad respects, which affect selection of topics as well as the technical level and manner in which they are treated.

1. *Trading agent focus.* Many discourses on auction theory emphasize the problem of the auction designer, highlighting such issues as relative revenue properties of alternative market mechanisms. The perspective here is exclusively that of the trading agent designer, whose problem is to participate in an existing market regardless of how or why it came to be. We engage in

comparative mechanism analysis to highlight implications of rule differences for trading agent strategy, but avoid distracting attention to questions about market design per se.

2. *Practical strategic reasoning.* Where other introductions to auction theory cater to those who aspire to extend the theory in new directions, the aim here is to generate lessons for constructing strategies in analogous situations. Accordingly, the exposition relies heavily on worked-out numeric examples, as opposed to formally stating and demonstrating abstract propositions. Our premise is that often the best way to understand the import of a strategic issue is to work it out in a concrete instance, notwithstanding the simplifications invoked. As will become apparent to the reader, even simplified auction scenarios can prove fairly complicated to work through in detail.

The lecture presupposes no prior background in game theory or auction theory, or artificial intelligence, although no doubt those with some prior exposure to these fields will find some of this material easier to get through. I make no attempt to systematically cover game theory, particularly as that topic is nicely treated by another *Synthesis Lecture* [Leyton-Brown and Shoham, 2008]. Game-theoretic concepts are rather introduced as needed, and applied going forward. The mathematics employed in this lecture lie mainly within algebra and probability, and a little bit of calculus. The reader with less mathematical inclination or energy can gloss over most derivations with measured loss in insight. However, I would urge those with fortitude to work through the example scenarios, as there is no substitute for actively solving a detailed problem for making a concept really sink in.

Except for an early chapter that motivates strategic issues through an extended case study of bidding on eBay, the lecture emphasizes generic market structures rather than specific markets or trading domains. In particular, the lecture contains no specific treatment of electronic trading in financial markets, despite the preponderance of trading activity devoted to financial securities. Similarly, those looking for dedicated case studies on electricity trading or Internet advertising or other vertical market will not find them here. Nevertheless, agent designers for any of these domains will find that many of the more abstract strategic lessons are directly applicable to their problems. For example, the double auction mechanism studied in Chapter 4 is an abstract version of the market institution employed almost universally by commodity and stock exchanges. Emphasis of issues such as bidding in simultaneous interdependent markets (Chapter 5) is fundamentally motivated by the relevance of these problems to real trading domains like electricity and keyword search advertising. No doubt specialized knowledge would inform trading agent design for particular applications, but limitations in space and time and expertise preclude detailed treatments here.

Writing this lecture was certainly a great learning experience for its author. The content draws significantly on collaborative research experiences with students and colleagues, especially Amy Greenwald, Chris Kiekintveld, Jeff MacKie-Mason, Anya Osepayshvili, Dan Reeves, and Julian Schvartzman. Along with other members of the Strategic Reasoning Group at the University of Michigan Artificial Intelligence Laboratory over the years, they deserve much credit for shaping the perspectives and ideas I attempt to convey here. I am particularly grateful to individuals who provided constructive feedback on earlier drafts, namely Ben Cassell, Quang Duong, Amy Greenwald, Erika Homann, Kevin Leyton-Brown, Peter Stone, Prateek Tandon, and Bryce Wiedenbeck. You are fortunate to have been spared the errors and confusions they uncovered, if not those I kept hidden until after their reviews.

This book is dedicated to Julian and Clara, my favorite autonomous agents.

Michael P. Wellman
Ann Arbor, Michigan
July 2011

CHAPTER 1

Introduction

This lecture is a guide to the design and analysis of trading agents. Before diving in, let us discuss what trading agents are, why they are important, and what you may hope to learn from this lecture.

1.1 WHAT IS A TRADING AGENT?

Simply stated, a *trading agent* is an automated strategy for trading in electronic markets. *Trading* refers to activity that leads to exchange of goods or services for money. The word *agent* signifies the autonomy of the software program implementing the strategy—that it operates with limited human intervention. As Russell and Norvig [2009] put it, an agent is autonomous to the extent it relies on its own percepts (observations), rather than the prior knowledge of the designer. Thus, a program that simply relays orders to the market would not merit the designation "trading agent". We require at minimum that the automated strategy be expressed conditional on market information. This definition admits agents operating on a wide spectrum of degrees of autonomy, from simple rule execution to complex algorithms that employ sophisticated machine learning or strategic reasoning techniques.

Trading agents have existed as long as we have had computer networks and standardized interfaces to market institutions. Such automation of financial markets started emerging in the 1970s, and gained significant momentum through the 1980s. The explosion of the World-Wide Web in the 1990s opened new vistas for market automation, including consumer-oriented auction sites like eBay, as well as a plethora of business-to-business marketplaces. One of the most fertile areas for growth in automated trading today is in Internet advertising, for example in markets for the placement of ads alongside the results of a keyword search. In effect, every web search query or sponsored page-view triggers an auction, allocating ad slots to the best bids (commonly placed by trading agents). In these and other areas throughout commerce and finance, design and deployment of electronic markets remains an active area of innovation, by academic researchers, public policy makers, and entrepreneurial firms.

Computerization of trading environments has thus played an enabling role for trading agents. Advances in economic knowledge about trading strategy have also provided a major impetus behind the automation of trading function. For example, almost immediately after the Black-Scholes formula for valuing options was published in 1973, algorithms incorporating the formula on mainframes or even programmable calculators became an indispensable tool for option traders (whether or not the trades themselves could be executed electronically). Similarly, market-making strategies based on inventory control theory have been directly exploited by securities dealers. More fundamentally, *auction theory*—introduced by Vickrey [1961] and developed into a rich body of knowledge by economists in the ensuing half century—has provided a general framework for strategic analysis of market institutions. The machinery of auction theory has had significant influence in the design of market institutions as well as trading strategies, and underlies the perspective taken in this volume.

A third major technical domain driving the development of trading agents is artificial intelligence (AI), particularly through techniques for decision-theoretic reasoning and machine learning. Although specific automated strategies employed by financial trading firms are closely guarded secrets, their widespread reliance on methods from AI and machine learning is well known. Machine learning automates the generation of models from data, for example models that predict future prices or other values that bear on the profitability of potential trading activity. The field of AI more generally is concerned with design of autonomous systems, including computational techniques and development methodologies that promote construction of agents that effectively and reliably promote the interests of their designers.

The level of effort by practitioners devoted to trading applications is unsurprising, given the direct returns to successful endeavors. From the perspective of research on autonomous agents, the trading domain has the additional advantage of clear evaluation measures, and ubiquitous interest. Equally attractive for researchers is the challenge of dealing with complex strategic interactions, without the distractions of other forms of complexity posed, for example, by physical environments or natural language interfaces (interesting challenges in their own right, of course).

1.2   BARGAINING AND TRADING

The problem of trading is closely related to that of *bargaining* or *negotiation*. In a bargaining situation, according to Muthoo [1999], "players can mutually benefit from reaching agreement…, but have conflicting interests". Such is certainly the case in trading scenarios. By definition, a trading scenario features the prospect of a mutually beneficial exchange (presuming trades are voluntary), and buyers and sellers inherently have conflicting interests: on price, in particular. Many concepts central to bargaining—making and accepting offers, considering outside options, and asymmetric information, for example—are likewise crucial to the description and understanding of trading situations. The theoretical machinery of both bargaining and markets (auctions) share a foundation in microeconomics and game theory.

Yet the trading situations of concern here do not exactly coincide with the standard notion of bargaining situations. They are more specific, in that the agreements in question are restricted to exchanges of goods and services for money. The particular form of agreement enables us to impose structure on the bargaining process underlying trading. Specifically, trading activity is mediated by what we term *market mechanisms* (defined more precisely in Section 3.1), which govern the rules by which potential traders interact. Trading strategies therefore are explicitly tailored to market mechanisms. At the same time, trading scenarios are broader than the typical bargaining problem formulation, which focuses on two parties conducting a bilateral negotiation. Whereas multilateral bargaining is conceivable, in the trading context having multiple parties on one or both sides of the market is the norm. For this reason, research on automated negotiation and bargaining is relevant, but not always directly applicable to trading agent design and analysis.

1.3   ARBITRAGE AGENTS

The power of automating trading strategy is perhaps best illustrated by the example of agents designed to exploit *arbitrage* opportunities. Catching opportunities for certain profit is a basic capability for trading agents, and thus an appropriate starting point for our discussion of trading agent design.

From the *Dictionary of Modern Economics* [Pearce, 1983]:

> **Arbitrage.** An operation involving simultaneous purchases and sales of an asset, e.g., a commodity or currency, in two or more markets between which there are price differences or discrepancies. The arbitrageur aims to profit from the price difference; the effect of his actions is to lessen or eliminate it.

For example, if a good is exchanged on two markets, A and B, and the price that we can get for the good at A is more than it would cost to purchase at B, we can make a certain profit by buying at B and selling at A. Before jumping at the opportunity, however, we should consider several possible impediments to pocketing the price difference:

1. *transaction costs* for executing the trades;

2. *transport costs* (if the good is physical and the markets have different locations); and

3. *execution risk*: will the quoted prices persist over the time it takes us to carry out the trade?

We can account for costs straightforwardly, by simply requiring that the trades be profitable net of such costs. For financial securities, where transaction costs are small and transport irrelevant (although there may be some analogous processes to transfer a security from one administrative domain to another), the upshot will be to impose a threshold gross profit margin for triggering the arbitrage operation. Execution risk is a factor whenever there is latency between price information and the trade initiation (i.e., virtually always). This may be addressed roughly through incremental padding of the margin threshold; a more principled approach would entail assessing the probabilities and consequences of short-term price movements that threaten the arbitrage situation.

The identity relation at the heart of an arbitrage opportunity need not be so direct. For example, arbitrage in currency trading may involve chains of exchange. Let $x_{\$\euro}$ denote the exchange rate between dollars and euros: the price for obtaining one euro is $x_{\$\euro}$ dollars. Similarly, $x_{\euro\yen}$ denotes the price of yen in euros, and $x_{\yen\$}$ the price of dollars in yen. If at any point $x_{\yen\$} > x_{\$\euro}x_{\euro\yen}$, an arbitrageur can profit by purchasing $x_{\euro\yen}$ euros for $x_{\$\euro}x_{\euro\yen}$ dollars. These euros are sufficient to buy one yen, which then can be sold at a profit for $x_{\yen\$}$ dollars. Of course, the qualifications about transaction costs (typically expressed as a spread between buy and sell exchange rates) and execution risk still apply. This reasoning can be extended to an arbitrary number of currencies related by pairwise exchange rates, and arbitrage opportunities can be identified in a computationally efficient manner by applying variants of shortest-path algorithms to the exchange-rate graph.

Futures markets may also enable arbitrage, in this case through an indirection across time. For example, consider a contract to deliver a good at a certain price one year from today. If the future price exceeds $1 + r$ times the current (spot) price, where r is the interest rate, then the arbitrageur can borrow funds to buy the good now, and deliver it at a profit in one year. As above, we must account for any extra costs, such as that of storing the good pending delivery.

One of the most pervasive forms of arbitrage is based on *index securities*. An index security is defined by equivalence to a portfolio of underlying assets. Whenever its price varies sufficiently from the sum (weighted by composition of the portfolio) of the underlying asset prices, we have an arbitrage opportunity. Trading is particularly popular in index futures, in which case an arbitrage analysis must account for both the underlying security prices and the interest rate, as above. For stock index futures it is also necessary to factor in dividend payments [Hull, 2000].

Index arbitrage is a perfect example where strategy automation is necessary and sufficient for effective performance. It is necessary because the valuation for an index over a portfolio and with interest and dividend adjustments is too complex for rapid and reliable human calculation. Moreover, executing a simultaneous trade of the index contract and all of the underlying securities requires automated order submission through electronic market interfaces to achieve the speed associated with acceptable levels of execution risk. Arbitrage agents are sufficient because the formula for index pricing is straightforward, and computationally simple given the prices of underlying securities. Indeed, index arbitrage accounts for a significant segment of trading on US equity markets. The New York Stock Exchange reports the fraction of transaction volume attributed to program trading, which it defines as transactions involving 15 or more simultaneous securities and a minimum dollar volume. The fraction of program trading volume is regularly well over half, and much of that is attributable to index arbitrage specifically.

As long as there is some identity relation among assets that can be defined clearly, a strategy for trading on deviations from the identity can be implemented by arbitrage agents. The problem of searching for deviations is ideally suited to automation, as computers are adept at monitoring large streams of data and verifying well-specified conditions. As suggested by the definition of arbitrage above, however, this very adeptness conspires to render arbitrage opportunities slight and scarce. Computers are cheap, and competition among arbitrage agents drives the price (i.e., the margin threshold triggering profit taking) down. [1] In consequence, the leading edge in arbitrage strategy is to discover yet more indirect identity relationships, and to stretch the very notion of identity. Strategies based on relationships that hold probabilistically (as induced by historical observation) but not by definition are termed *statistical arbitrage*, and represent a major category of automated trading strategy in financial markets. For example, in the *pairs trading* technique, the arbitrageur identifies two securities that tend to move in tandem—say, shares of Coca Cola and Pepsico—and

---

[1]Trading firms also compete on speed, leading to a latency-reduction arms race (one of the factors driving *high-frequency trading*) based on high-performance hardware and software and placement of machines as close as possible to the trading floor. This dimension is orthogonal to strategy, and so not discussed further here.

exploits temporary deviations from their historic price relationship by buying one and selling the other.

1.4   THIS LECTURE

Our discussion of arbitrage comes in the Introduction chapter because it represents a conceptually straightforward task for trading agents, and often the first objective for automating trading function. In the chapters below I take the capability to detect arbitrage opportunities for granted, and focus on trading objectives that remain once arbitrage relations are fully exploited. That is, we consider agents who exchange not for pure trade profit, but in order to satisfy needs for consumption, or to operate an economic production process. These fundamental values are what makes trading ultimately beneficial to engaged parties, and essential for functioning economies.

In presenting the principles of trading agent design and analysis, I appeal primarily to knowledge from the disciplines of economics and artificial intelligence. The treatment draws specifically from auction theory, techniques for strategic reasoning, and empirical studies of real and simulated trading environments. Throughout, I emphasize the implications of insights from these fields for trading strategy. Although the sequel asserts many general results, with technical arguments and derivations, I sacrifice formal theorem statements and proofs to elevate the conceptual insights, and convey the essential logic of the strategic reasoning involved.

We start in Chapter 2 by methodically exploring a familiar example: bidding in eBay auctions. Thorough analysis of a simple case conveys some strategic insights without requiring the formal terminology and technical machinery employed in the rest of this text. The eBay example motivates careful attention to details of a trading environment, and establishes some basic patterns of strategic thinking we exercise throughout.

Chapter 3 presents the fundamental technical characterization of auctions and models of bidder objectives necessary for a theoretically grounded analysis of trading environments. This introduction to auction theory surveys some basic auction types and valuation models, which serve as building blocks for a wide range of market scenarios, including the more complex auction mechanisms considered in subsequent chapters. The treatment here maintains the perspective of the trading agent—as opposed to that of the auction designer taken by the majority of such surveys.

Chapter 4 focuses on a particular market mechanism, the continuous double auction (CDA). CDAs are at the core of almost all financial and commodity markets, and as such they drive the vast majority of automated trading activity. Despite their ubiquity and simplicity of rules, however, the dynamic nature of CDAs renders their strategic analysis theoretically difficult. Our treatment of CDA bidding in this chapter demonstrates the value of agent-based simulation for evaluating strategy ideas.

Chapter 5 deals with strategies for trading in multiple simultaneous markets. Whereas ability to monitor activity in many markets at once is a strength of trading agents, handling the interactions across markets is also a significant challenge. Techniques based on price prediction and marginal value assessment provide a general way to manage market interdependencies, and can be incorporated directly into trading strategies. The analysis in this chapter combines the theoretical and empirical approaches, demonstrating their synergistic benefits for gaining a strategic handle on complex market domains.

The final chapter reviews the major lessons distilled from the foregoing analyses. By way of summary, I sketch elements of a general architecture for trading agents, and illustrate how the steps of a generic bidding cycle can be realized in a trading agent for a complex supply chain scenario. The volume concludes with an assessment of our state of knowledge about trading agent strategy, and speculations on the future of trading agents, in research and practice.

1.5 BIBLIOGRAPHIC NOTES

To avoid unduly cluttering up the main exposition with citations, end-of-chapter bibliographic sections provide original sources for key points and resources for further reading. For this Introduction, the focus of the Notes is on topics not covered in depth elsewhere in this volume.

Domowitz [1993] documents the early development of automated financial markets. Kim [2007] provides a guide to more recent state of art. Miller [2002] discusses many interesting issues for strategy in financial markets, including implications of options pricing models. Market-making algorithms employed by securities dealers is the focus of the theory of *market microstructure* [Garman, 1976, O'Hara, 1995]. Gatev et al. [2006] describe and assess the statistical arbitrage method of pairs trading. The prevalence of techniques from AI and machine learning in algorithmic trading is widely reported in press features [Duhigg, 2006, Salmon and Stokes, 2010] and other popular accounts [Bass, 1999].

Auction theory as a field can be defined as the application of game theory to market interactions. It is a mature and active subject of inquiry in economics [Krishna, 2010], which we delve into further (although far from exhaustively) in Chapter 3. Artificial intelligence is likewise a long-established field with an extensive literature. The text by Russell and Norvig [2009] provides a comprehensive treatment of the field's central concepts. Machine learning is a specialization within AI, represented by numerous texts [Bishop, 2006, Hastie et al., 2001, Szepesvári, 2010] emphasizing different thrusts within the field, and a large and active research literature.

Another AI subfield, *multiagent systems* (MAS), focuses on issues that arise from the interaction of multiple autonomous agents. MAS includes an active community of researchers devoted to negotiation and bargaining agents [Kraus, 2001, Sycara and Dai, 2010]. The text by Shoham and Leyton-Brown [2009] reflects the increasing influence of economic and game-theoretic approaches in MAS research. Also within MAS, a significant literature (a small fraction represented by the bibliography here) addresses problems posed by trading domains. Some of this research is presented at the *Workshop on Trading Agent Design and Analysis*, which is held in conjunction with an annual Trading Agent Competition (TAC) tournament event. [2] TAC has spurred numerous advances in trading agent research [Wellman et al., 2007], including many of the ideas presented in Chapter 5.

---

[2]http://tradingagents.org

Lahaie et al.   [2007]   describe   how   sponsored-search   keyword   auctions   work, and Kitts and Leblanc [2004] some of the pioneering work on automated trading strategies for this domain. A TAC game devoted to ad auctions was introduced in 2009 [Jordan et al., 2010], and trading agents showcasing sophisticated strategies for this domain appeared immediately [Pardoe et al., 2010].

CHAPTER 2

Example: Bidding on eBay

We begin by examining a simple and familiar trading setting: buying goods at online auctions such as eBay. The simplicity of eBay bidding allows us to thoroughly analyze this case, and by examining the situation in detail we can observe some subtle strategic issues.

eBay (http://www.ebay.com) is by far the most popular Internet auction site, listing 113 million items concurrently, with $2,000 of goods traded every second.[1] Started in 1995 by Pierre Omidyar, eBay's genesis and growth into a significant forum for commerce is a classic success story of the early Internet economy [Cohen, 2002]. As of September 2009, eBay reported that on its site, on average, a pair of shoes was sold every four seconds, a major appliance every minute, and a cell phone every five seconds.

2.1 EBAY AUCTION RULES

The first step in designing a trading strategy for a given market environment is to understand thoroughly the rules governing the bidding process. On eBay, goods are sold in auctions,[2] each initiated by a seller listing its [3] good for sale. The listing specifies the good, typically with descriptive text accompanied by one or more pictures. It also discloses terms of the transaction, including shipping costs, return policy, and accepted modes of payment. The eBay identifier of the listing seller is provided, from which one can obtain information such as the seller's feedback score on eBay's reputation mechanism. Finally, the listing specifies parameters of the auction, namely the end time and starting or current bid price, and details such as whether the seller has imposed a secret reserve price below which it will not sell the good.

An eBay auction is a variation on one of the most familiar auction types, known technically as the *English open-outcry auction* (defined in Section 3.2.4 below). Stated most simply, eBay buyers submit bids specifying the price they are willing to pay for the item. In order to be admissible, a bid must name a price greater than that of the current bid (henceforth, BID), plus a standard increment δ that depends on the current price range. For example, if $BID = \$11.50$, then $\delta = \$0.50$, and so a new offer b must be for at least $ASK = BID + \delta = \$12.00$. The values BID and ASK are called *price quotes*. We define these more generally below, but for now we can think of BID as representing the current selling price, and ASK as the price one would have to offer to become the winning buyer.

---

[1] Source: eBay Marketplaces Fact Sheet, Q4 2008.
[2] eBay also supports fixed-price offerings, which they term *BuyItNow* listings. In this discussion, we focus on auctions, where prices are determined dynamically by the market.
[3] We employ the neutral gender throughout for traders of unspecified sex, including software agents.

On eBay, ASK defines the minimum admissible offer price, but bidders may offer strictly more. Under eBay's *proxy bidding* system, an offer b represents the maximum the bidder is willing to pay, but the bidder will not actually pay that much unless necessary to beat out other bidders. Here is how eBay explains it:[4]

1. When you place a bid, you enter the maximum amount you're willing to pay for the item. The seller and other bidders won't know your maximum bid amount.

2. eBay places bids on your behalf starting with the next bid increment for the auction. We'll bid as much as necessary to make sure that you remain the high bidder (or to meet the reserve price). We'll keep bidding for you until bidding reaches your maximum amount.

3. If another bidder has placed the same bid before you or a higher maximum bid, we'll let you know that you've been outbid so that you can place another bid if you want. However, if no other bidder has a higher maximum bid at the end of the auction, you win the item even if your bid doesn't go as high as your maximum bid. You could pay significantly less than your maximum price. This means you don't have to keep coming back to re-bid every time another bid is placed.

In other words, submitting a bid $b \geq ASK$ can be viewed as two separate acts. The first act submits an offer to the auction at price ASK. This then becomes the new current bid price, BID. The second act authorizes eBay to submit subsequent bids on the bidder's behalf (i.e., as *proxy*), as needed, up to the amount b. At the end of the auction, whichever bid has last been admitted to the auction (submitted directly by the bidder or by eBay as proxy) determines the winner, and the final value of BID the winning price.

An alternative way to define the auction is as follows. The auction maintains a current bid price, BID (initially zero), and a value b^* representing the highest bid submitted thus far. The value BID is published, whereas b^* remains hidden. ASK is defined as $BID + \delta$, where the increment δ may depend on the applicable bid level, and is initially set to the starting ASK price. When a new bid $b \geq ASK$ is admitted, these values are updated as follows. We distinguish two cases. First, if the new bid exceeds the highest submitted thus far, $b > b^*$:

$$BID \leftarrow \min(b, b^* + \delta),$$
$$b^* \leftarrow b.$$

Otherwise, the new bid exceeds ASK but not the current b^*:

$$BID \leftarrow \min(b^*, b + \delta),$$

and b^* is unchanged. At the end of the auction, the winner is whomever submitted the bid that last updated b^*, and the final value of BID is the winning price.

[4]http://pages.ebay.com/help/buy/proxy-bidding.html, July 2008.

These descriptions are equivalent in the sense that their outcomes (final winners and prices) are the same, as are all observable states (price quotes), for any sequence of user bids. The former description, based on proxy bidding, may be especially easy to understand for people familiar with stereotypical English auctions. The latter, more precise description is presented in terms more suggestive of a *second-price auction* (defined and analyzed in Section 3.2.2), and is often more convenient for implementation and strategic analysis.

To test your understanding of the bidding rules, view the outcome pages of some completed eBay auctions, and attempt to reconstruct the series of bids that produced this result. An example bid history is displayed in Table 2.1. Twelve bidders submitted 19 bids over a three-day period, competing to obtain a beautiful accordion made by La Burdina in Italy.

In interpreting Table 2.1, note that bids are listed by amount, not chronologically. The auction starts with $BID = 0$, $ASK = 0.99$, and $b^* = 0$. The first admitted bid is submitted about three

Table 2.1: An example bid history from an eBay auction in February 2010. Bidder a***n won the accordion with a bid of at least US$682.99, submitted just two seconds before the auction ended.

Bidder	Bid Amount (US$)	Bid Time
a***n	**682.99**	Feb-13-10 18:54:32
r***5	*672.99*	Feb-13-10 17:27:32
s***a	*600.00*	Feb-12-10 14:37:25
b***5	460.00	Feb-12-10 16:31:24
e***e	*450.00*	Feb-12-10 07:43:10
t***t	*426.99*	Feb-11-10 05:14:17
e***e	420.00	Feb-12-10 07:42:56
e***e	400.00	Feb-12-10 07:42:26
e***e	350.00	Feb-12-10 07:42:13
e***e	*320.00*	Feb-12-10 07:41:49
0***2	285.00	Feb-11-10 06:16:33
0***2	275.00	Feb-11-10 06:15:58
0***2	265.00	Feb-11-10 06:15:30
n***e	*252.52*	Feb-11-10 03:07:47
i***u	60.00	Feb-11-10 03:44:32
i***u	52.00	Feb-11-10 03:44:15
a***p	*50.00*	Feb-10-10 21:14:40
u***l	*22.00*	Feb-10-10 18:57:28
s***d	1.04	Feb-10-10 20:37:48
starting ASK	0.99	Feb-10-10 18:54:34

minutes after auction start by bidder u***l (eBay obscures the actual bidder identities in this way; I find it useful to read the name as "U stars L"). The table shows this amount in italics, indicating that it represents a high bid at the time of submission. At this point, b^* is updated to 22, but this amount is hidden from bidders. BID is updated to 0.99. The next bidder, s***d, offers \$1.04. This is not nearly enough to beat u***l's proxy, but does trigger the update $BID \leftarrow 1.04$. A short time later, a***p submits a bid of \$50. The applicable increment at $b^* = 22$ is $\delta = 0.5$, thus the new bid produces $BID \leftarrow 22.50$ and $b^* \leftarrow 50$. The next bidder, n***e, offers \$252.52 for the accordion. Updating the auction state and tracing the remaining bids is left as an exercise for the reader.

Now we turn to the real question of interest: How should we design a software agent to bid on eBay on our behalf? The core of this problem is specifying the *bidding strategy*: what bids the agent should submit and when, as a function of our given objectives and any information available to the agent during the auction. Suppose, for example, we are interested in purchasing an accordion (perhaps not as fine as the example above) we find available in a current eBay auction. Upon consideration, we estimate that the accordion has a value to us of \$100, which means that we would expect to be equally happy to get the instrument and pay \$100, or to not get it but keep the money. [5]

We focus on three strategic issues bearing on how we bid for this accordion. The first is a question of timing—should we bid right away, or wait until near the end of the auction? The second issue is deciding at what price to bid. The third is a question of shopping—how does the potential availability of other opportunities to buy accordions, on eBay or elsewhere, affect how we should bid in the current auction? The following sections address these questions, in turn.

2.2   TIMING

Suppose we have decided to bid \$90 for the accordion we are trying to purchase. Let us defer discussion of whether this amount is appropriate to the next section. Say the current price (BID) is \$50, and the auction is scheduled to end in 48 hours. Should we enter our \$90 bid now, or wait?

At first glance, the proxy bidding mechanism may appear to neutralize the choice. Entering a bid value of \$90 expresses the *maximum* we might have to pay; if our bid is highest the actual amount we pay is only as much as necessary to beat the others. For a fixed set of lower bids by other agents, in fact, it is virtually *irrelevant* when we enter ours, as long as it is before the deadline. [6]

The catch in this analysis is that the other bids should not be viewed as fixed, since entering our bid of 90 could affect the decisions of other bidders. How might this happen? Consider the example auction history displayed in Table 2.1, with particular attention to the behavior of bidder i***u. This bidder first offers \$52 in a state with $BID = 51$, which is not sufficient to exceed b^* as

---

[5]Making such value judgments—the *preference elicitation* problem—is itself a challenging element of market decision-making, the subject of much research in decision analysis and artificial intelligence. For our present purposes we ignore the subtle issues involved, and take for granted the ability to render such value judgments. Our key assumption is that values can be represented on a monetary scale, and specified independently of other market decisions. We relax the latter assumption in our treatment of more complex market environments (Chapter 5).

[6]For the case of an exact tie, the earlier bid wins. Furthermore, if the current price reaches a point between $90 - \delta$ and 90, we would no longer be allowed to submit a bid of 90, even though no other bid is greater. These are two tiny advantages to early bidding.

set by the prior proxy bid by n***e. Within a minute of finding out that what it just entered was insufficient to win, i***u increases its bid to $60. Although we cannot be sure what this bidder would have done if n***e's bid did not exist, a reasonable hypothesis is that i***u would have been satisfied to stay at its original bid if eBay had identified this as the current high bid. i***u's full strategy is unobservable, but we can rule out that it bids its maximum willingness to pay from the beginning, as this is contradicted by its manifest behavior. An even more extreme example of this is exhibited by bidder e***e the next day. For whatever reason, several bidders in this auction offered considerably less originally than what they later revealed themselves willing to pay for this accordion.

How does the presence of *incremental bidders* like i***u and e***e affect our choice of bidding strategy? Imagine for the moment that all other bidders behave in this way. If we place our bid of $90 at the beginning, then we will win the good only if none of the incremental bidders are willing to offer this much. Our ultimate price will be the value at which the next-highest incremental bidder chose to drop out. Suppose instead we delay our bid, until the very end of the auction. Before we enter the fray, the incremental bidders will compete with each other, culminating in a situation where the other bidder with greatest willingness to pay (call it I*B) is the high bidder, and the current bid b' as entered by I*B is just enough to beat out the second-highest I*B-like bidder. This value could be considerably less than I*B would ultimately be willing to pay. If $b' > 90$, we will not bid and it turns out not to have mattered that we delayed our entry. If $b' + \delta \leq 90$, however, we can place our bid in the last few seconds of the auction and win (at a price of $b' + \delta$), since by the time I*B finds out it is outbid the auction will end and it will be too late to increase its offer. This strategy of placing a bid at the last moment—too late for others to respond—is called *sniping*, and is in fact a common practice in eBay bidding. In the example above, bidder a***n won the La Burdina accordion with a snipe.

In a landmark study of eBay bidding patterns, Roth and Ockenfels [2002] found that indeed a large fraction of bids are submitted near the auction deadline, and suggested that countering incremental bidders could be one good reason. Lee and Malmendier [2011] also found extensive incremental bidding for some items on eBay, including a surprising prevalence of bidders who continue to raise their bids beyond the fixed price at which the same item is simultaneously available on eBay. The authors speculate that such bidders get caught up in the excitement of bidding, and forget about their alternative options for purchasing the item. Numerous other studies have confirmed the prevalence of sniping on eBay, and Wintr [2008] observed that the tendency is considerably stronger in auctions where multiple bids by the same bidder (as for incremental bidding) appear. In comparison, data gleaned from auctions run by Amazon exhibited a much lower degree of last-moment bidding. This can be explained by a subtle difference in closing rules for eBay versus Amazon auctions. On Amazon, auctions close when their specified deadline is past *and* no bids have been received for ten minutes. Given this buffer for inactivity, sniping on Amazon auctions is strictly impossible, since incremental bidders always have at least ten minutes to respond to being outbid.

The phenomenon of incremental bidding represents one way that our bid can affect others' bids. Another is through transmission of information about the good being auctioned. How different bidders value this auction depends on many factors, some very specific to individual preferences, but other factors that may be shared by many parties. For example, the value of an accordion depends on its craftsmanship and condition, which may be difficult to assess from the auction description. More generally, bidders will typically share a common interest in an item's authenticity, rarity, or any factor affecting potential resale value. Suppose I am known to be an expert on assessing accordion quality. The fact that I am willing to bid on this particular instrument in effect reveals information to other bidders, which may lead them to reassess their own bids. When my bid reflects positively on the good, this has the potential to cause another bidder to outbid my own offer, or raise the price at which I could otherwise obtain the good.

The shared component of good valuation provides another strong incentive for me to hide my bid. I could adopt an obscure identity, [7] but even if my expertise is not known, the fact that *somebody* values the good so much nevertheless conveys positive information, counter to my interests. The only way to completely hide the information in my bid (until it is too late) is, once again, by sniping. As evidence for this effect, Roth and Ockenfels [2002] showed that a category of goods with a high presumed expertise component (antiques) exhibited a higher prevalence of late bidding than a category where values were widely known (computers). Wintr [2008] did not find this difference between categories, but confirmed the effect within the antiques category by comparing the timing of bids by identified experts with the relatively earlier bidding by collectors.

2.3   PRICING THE OFFER

Now that we have established *when* to bid, we can address the central question of *how much*. Was the $90 we chose to bid the right amount? A thorough treatment of this question requires some of the machinery we introduce in subsequent chapters. For now, we merely identify some conditions under which the answer is clear.

Recall our working assumption that we value the accordion at $100. If in addition:

• this value does not depend on the valuation of any other bidder;

• there will be no further opportunities to obtain this good or anything that could substitute for it;

• we submit our bid at the last moment (i.e., snipe),

then in fact $90 is *not* the right amount to bid, but rather we should bid at our value of $100.

These are stringent conditions, but they allow us to make a strong case for bidding at our valuation. The independence of other bidder values means that regardless of what we find out in the auction, our value for the good remains $100. The absence of related market opportunities (the *one-shot auction* assumption) means that the outcome of losing the auction removes forever the potential $100 value of obtaining this good. Since we are sniping, our bid will not influence any other bids.

---

[7] Actually, under eBay's current rules, bidder identities are hidden during and after the auction. This was not the case when the studies cited here were conducted.

We win the auction if and only if (iff) the highest bid submitted by others, b', satisfies $b' + \delta \leq 100$, in which case we pay $b' + \delta$. We are naturally happy to win at any such price, and given the other bids, no other value we might bid would get us a price better than $b' + \delta$. Bidding greater than \$100, moreover, would risk that we win the good and pay more than our value, which is clearly undesirable. Thus, under the conditions stated, bidding our true value is an optimal offer.

If bidding right at our value for the good seems counterintuitive, it could be because in practical experience the one-shot auction assumption rarely holds strictly. As we see below, the existence of alternative markets for the good could justify a bid of \$90, or other amount below our valuation.

2.4 SHOPPING

The final issue we address in our eBay case study considers situations where there are in fact future opportunities to purchase the good, beyond the current auction. This is indeed the norm for goods available at eBay, or in most trading forums for that matter. For example, at this moment, eBay lists 32 Bonetti brand accordions for sale, in various models and colors.

To take a simple case, suppose that we wish to buy a used accordion today, and there are two suitable opportunities: an eBay auction ending this morning, and another ending this afternoon. (We care only about the ending time since we intend to snipe.) Per our reasoning above, if the morning auction were our only chance to obtain the good, we would snipe for \$100. How does the existence of the afternoon auction change our strategy? It turns out that in analyzing bidding strategy over time, it is helpful to start at the end, and consider what we would do in the afternoon, *conditional* on what happened in the morning. If we had won the morning auction, then we accomplished our goal and we take the afternoon off from shopping. If we had lost, then our situation in the afternoon is like the case analyzed above, where we have exactly one opportunity. In fact, the reasoning of Section 2.3 applies, and we should bid \$100 in the afternoon if we lose in the morning.

Now, our decision about how much to bid in the morning can take account of this next-stage strategy. Intuitively, the existence of the afternoon opportunity has positive value in the case where we lose the morning auction. This value depends on the outcome of the afternoon auction—who wins and at what price, which is of course uncertain at the time we contemplate our morning bid. We represent uncertain beliefs using probability distributions. Specifically, let $G_{100}(p)$ denote the probability that we win and pay price no greater than p, given that we lost in the morning and bid \$100 in the afternoon. The *expected value* of the opportunity is determined by this probability, and our preferences over uncertain prospects for profit. We assume *risk neutral* preferences, which means that utility for outcomes is linear on the monetary scale. [8] The expected value v_A of bidding in the

---

[8] Risk neutrality is a reasonable model for relatively small decisions like buying an accordion. For situations involving considerably larger economic or life decisions, *risk aversion* or other nonlinear utility forms may be warranted, and could be accommodated at some increase in complexity of the reasoning. Here we generally assume risk-neutral preferences for simplicity, unless stated otherwise.

afternoon can then be expressed as

$$v_A = \int_0^{100} (100 - p) dG_{100}(p).$$

This corresponds to the value of the situation where we lose in the morning, since at that point all we have is the afternoon opportunity. In our original analysis of the one-shot case, losing the auction had zero value, so winning the good was $100 better than losing (not counting the price we pay for the win). Taking into account the afternoon recourse, winning is only $100 − v_A better. This, then, is the amount we should be willing to bid in the morning auction. At prices of $100 − v_A or less, we obtain more profit than we expect from the afternoon alternative, whereas at prices above $100 − v_A we are better off trying the later auction.

In addition to the assumptions already mentioned, this conclusion requires that our predictions about the afternoon auction (and thus its expected value v_A) do not depend on the outcome of the morning auction. In particular, the fact that we lost while bidding $100 − v_A does not alter our assessment of expected value v_A for bidding $100 in the afternoon. Given this form of independence, we could extend our reasoning to a longer series of sequential auctions. Having more future opportunities causes us to shade our bids lower, all else equal. A similar qualitative result would hold if the goods are not identical, or even if there were some dependence between earlier outcomes and later predictions. In such cases, the analysis would become substantially more complicated, and our later bids may be conditional on price outcomes from earlier auctions. We defer consideration of such complex scenarios to subsequent chapters, following introduction of some more technical machinery.

2.5   STRATEGY LESSONS

No site has done more than eBay to extend the practice of dynamic pricing through bidding to everyday commerce. Despite their simplicity, eBay auctions present many of the strategic issues facing bidders in auctions more generally, and help us illustrate some key principles underlying trading agent design. Thanks to its accessibility and commercial success, eBay has proved a popular forum for field research on auction behavior, leading to an extensive literature surveyed by Ockenfels et al. [2006].

Our case study serves to introduce some specific observations and prescriptions for bidding on eBay. For elaboration on these, and exploration of many more interesting strategic phenomena exhibited on this popular auction site, Steiglitz [2007] offers an engaging narrative guide. From the perspective of trading agents in general, the eBay example also illustrates some broader lessons. Although simple and informal, our discussion yields several key insights that in fact hold across trading domains:

1. *Auction rules matter.* As an example we saw that whether the auction closes at a fixed time or also requires a period of inactivity significantly influences the prevalence of sniping. This is merely one instance of the more general principle that auction rules affect bidder behavior.

In designing a trading strategy we must carefully analyze the auction rules, and if we were designing an auction we would need to account for the behavior induced by rule choice.

2. *Modeling other bidders matters.* In the eBay case study we saw that the presence of incremental-bidding behavior provides a sufficient reason to adopt a sniping strategy. It is useful to model not only direct behavior of others, but also their underlying values and information states. For example, the effect of early bidding by experts hinges on how others will react to the information thus revealed. And predicting the expected value of future auction opportunities as part of a multi-stage analysis inevitably relies on a model of other bidders and their valuations.

3. *The context of related markets matters.* The optimal strategy for bidding in a one-shot auction is no longer ideal in a situation where related opportunities are available in subsequent or simultaneous auctions. More generally, a comprehensive trading strategy must deal with the entire configuration of relevant markets, not individual auction mechanisms, even when these are operated independently by different parties.

As we explore the range of trading scenarios in subsequent chapters, we revisit these lessons repeatedly, and consider their specific implications.

CHAPTER 3

Auction Fundamentals

This chapter presents the essential theoretical framework for analyzing strategic behavior in trading domains: *auction theory*, the application of game theory to markets. Precise definitions of auction mechanisms are prerequisite for formulating market situations in game-theoretic terms. Surveying the basic abstract auction types serves to illustrate this mode of strategic reasoning, and provides concepts and tools applicable to more complex (and realistic) mechanisms. Since the basic types serve as building blocks for these complex auctions, insights from the basic analyses often carry over to the richer settings as well.

Equally important for a game-theoretic treatment of auctions are models of the objectives of participating agents. As we see below, the source of agents' valuations for goods traded, and the information agents have about each others' valuations, is pivotal in the strategic analysis of auction scenarios.

3.1 AUCTIONS AND MARKETS

The term "auction" applies to any well-defined mechanism for determining the parties and terms of a market-based transaction. To make sense of this statement, we need to introduce some familiar concepts in an atypically precise manner.

A *market* is an environment where agents interact in order to exchange goods and services.[1] To facilitate multilateral exchange, such environments generally employ standardized currency, or *money*, which allow the terms of exchange for a variety of goods to be expressed with respect to a common scale of value. A *market-based transaction* is thus an exchange of goods for money. If the good also has a standardized description (i.e., so agents understand what it constitutes), then it can be associated with a *price* representing the quantity of money for which it can be exchanged.

Let us suppose that the standardized goods are numbered, with p_i the price of good i. If goods can be exchanged in quantity, then we often conceive of these prices as applying per unit.[2] Unit prices define the ratios at which goods can be effectively exchanged for each other. That is, at the going prices, a unit of good j could be traded for p_j/p_i units of good i. One of the virtues of market-based exchange is that we can achieve a complex multilateral exchange of goods among a large set of agents through a series of bilateral transactions, each involving one party trading goods to another for money. The alternative is *barter* (exchange of goods for goods), in which agents must

---

[1]Henceforth, for simplicity I employ the term *goods* to also encompass services.

[2]For the nonce, we assume that the unit price is constant with respect to quantity. This is without loss of generality, as we can always treat different quantities as representing different goods. Nevertheless, it will often be more convenient to avoid this proliferation of distinctions and instead allow varying unit prices, also called *nonlinear pricing*.

find exchange partners who have the goods they want, and want the goods they have. As long as agents in the market believe that prices are relatively stable, they will be willing to accept money for goods, knowing they can obtain goods they want for this money at a later time.

Where do market prices come from? Loosely speaking, prices are determined through the communications of agents offering to buy and sell goods to each other. Any actual transaction represents an agreement of agents to exchange, and the money per unit of good is the corresponding price. The terms of exchange may have been determined in a variety of ways. Determination processes that operate among agents according to well-defined rules are called *mechanisms*.

More formally, a *resource allocation mechanism* comprises the rules of an allocation process, specifying (1) permissible actions (often limited to messages, expressible as a communication protocol) and (2) allocation outcomes as a function of agent actions. In a *market mechanism*, the possible ultimate outcomes comprise market-based exchange transactions. A mechanism is *mediated* if there is some entity, distinct from the participants, that manages the communication and implements the mechanism rules.

Finally, we can define *auctions*, as mediated market mechanisms. That is, an auction is a mediated process, operating according to well-defined rules, that determines the terms of market-based exchanges. Auction rules define the communication possibilities for each participant (see Figure 3.1), including:

- *bids*: expressions of offers, where agents indicate some exchanges they are willing to undertake, and

- *quotes*: information sent by the auction to the agents, summarizing the state of the auction process (e.g., the current price).

The result of an auction specifies a set (possibly empty) of exchanges, each dictating which agent buys what good from which other agent, at what price. Auction rules define how this outcome is determined, as a function of the agent communications.

3.2   BASIC AUCTION TYPES

In the simplest allocation scenario, we have one unit of a single good, owned by a particular agent, which we label agent 0. Since agent 0 may be willing to exchange this good for money, we refer to this agent as the seller. Let there be N other agents, numbered $1, \ldots, N$, who are potential buyers.

An auction mechanism for this scenario would specify the communication rules:

- the form of offers to exchange the good, and when they are allowed, and

- the timing and content of quote information, if any.

The auction would also define its allocation rule: how it determines who (expressed as an agent index, j) gets the good, at what price (p). For the case of no sale, we say $j = 0$.

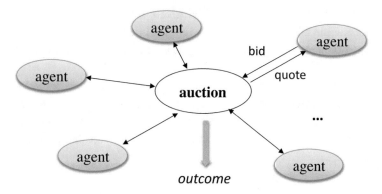

Figure 3.1: Communication patterns in a mediated market mechanism, or *auction*. Agents submit bids to the auction, and the auction disseminates intermediate price quotes and reports a final outcome.

Although the single-item auction is described in terms of one seller and many buyers, a simple transformation specifies a *procurement auction* with one buyer and many sellers. We simply reverse the terminology, and let j denote the agent who wins the right to sell the good at price p. Whereas higher-priced offers are preferred when the buyers bid, lower-priced offers prevail in the procurement case. All of the results and arguments below can be recast for the procurement setting through such a transformation of polarity. [3]

3.2.1   FIRST-PRICE SEALED BID

Possibly the most straightforward mechanism for this scenario is the *first-price sealed-bid* auction (FPSB). The form of a bid b_i for buyer i is simply a price that i is willing to pay for the good. The auction provides no price quotes, therefore no agent learns anything about the bids submitted by others until the auction is over. (This is of course what makes it "sealed-bid".) The allocation rule is simply that whichever buyer submits the highest bid ($\arg\max_i b_i$) buys the good from agent 0, at price b_i. (That the winning buyer pays the highest price offered is what makes it "first-price".)

How should an agent bid in the FPSB mechanism? Before we can start to answer this question, we need to define the agent's objective. Specifically, we define a *utility function*, $u_i(j, p)$, representing agent i's preference for possible outcomes of the auction. Presumably, an agent would consider buying the good at auction only because it would value having the good in its possession. We denote by v_i agent i's valuation for the good, and express its utility for auction results as follows:

$$u_i(j, p) = \begin{cases} v_i - p & \text{if } j = i \\ 0 & \text{otherwise.} \end{cases} \tag{3.1}$$

---

[3]Sometimes procurement auctions are termed *reverse auctions*. Since this label reinforces a notion that bidding to buy is the "normal" form of auction, we avoid such terminology here. Whether the mechanism admits bids to buy, to sell, or both—if it mediates market exchange, it's an auction.

The utility function (3.1) is called *quasilinear*, which means that the overall utility is an additive combination of the auction allocation and the payment. In essence, it entails that the good valuation is expressible on a monetary scale. We assume quasilinearity throughout our analysis of auction mechanisms.

We can now frame the strategy question more precisely. Agent i's strategy in general takes the form of a function over valuations, $b_i : \mathbb{R}^+ \to \mathbb{R}^+$. Let $\Pr_i(\text{win} \mid b, v_i)$ denote the probability that agent i wins the auction (i.e., gets the good), given that it has valuation v_i and bids an amount b. If it does win, it pays b, according to the FPSB rules. The agent's expected payoff given its valuation can thus be written

$$\Pr_i(\text{win} \mid b(v_i), v_i)[v_i - b(v_i)], \tag{3.2}$$

since i's utility is zero if it does not win the auction.

What price, $b_i(v_i)$, then, should agent i bid, given its valuation? First, it is immediate from the expression of expected payoff (3.2) that $b_i(v_i) \leq v_i$. Bidding more than its valuation risks achieving negative utility if the agent "wins" the good, whereas it can guarantee zero utility simply by bidding zero (equivalent to not bidding, since $\Pr_i(\text{win} \mid 0) = 0$). For the special case of zero valuation, nothing better is possible, so $b_i(0) = 0$. With positive valuation, however, the agent has some opportunity for gain. In that case, we can see that bidding exactly v_i cannot be optimal, since that strategy *always* achieves a utility of zero. Rather, the agent with positive valuation should invariably *shade* its bid downward, $0 \leq b_i(v_i) < v_i$, so that it achieves strictly positive utility in the case where it wins the good.

Exactly how much less than its valuation it should bid, however, depends on how the bid affects its probability of winning the good. Whether a given bid will win in turn depends on the bids of other agents, which are themselves governed by strategies expressible as functions of *their* valuations. To account for this factor, we require yet more precision in our specification of the decision environment. In particular, we need a model that addresses the valuations of all agents, including the uncertainty that each agent may have about the valuations of others.

Our starting point is the *independent private values* (IPV) model. Under IPV, each agent's valuation is characterized by a probability distribution, $F_i(v) = \Pr(v_i \leq v)$. The IPV model further assumes that agent valuations are probabilistically independent, leading to the joint probability $F(v^1, \ldots, v^N) = \prod_i F_i(v^i)$. Conceptually, the agents' valuations are initially uncertain, and drawn from this distribution before the auction. Given independence, observing one's own valuation provides no information about the valuations of others. It therefore says nothing about the bids of others, so we have $\Pr_i(\text{win} \mid b, v_i) = \Pr_i(\text{win} \mid b)$, resulting in the simplified expected payoff expression

$$\Pr_i(\text{win} \mid b(v_i))[v_i - b(v_i)]. \tag{3.3}$$

The IPV model also assumes that each agent has the same beliefs about others' valuations, modulo perfect information about its own. Moreover, each agent knows that the others' beliefs are the same, and that all others know this as well, *etc.*—a condition called *common knowledge*.

A consequence of independence is that an agent's strategy should be *monotone* in its valuation:

$$v > v' \implies b_i(v) \geq b_i(v').$$

To see this, consider the agent's objective, to maximize expected payoff (3.3). The corresponding first-order condition is given by

$$\Pr_i(\text{win} \mid b(v)) = [v - b(v)]\frac{d}{db(v)}\Pr_i(\text{win} \mid b(v)).$$

Since the probability of winning is nondecreasing in bid, the derivative term is positive. Increasing v thus requires a compensating increase in bid, to maintain the optimum.

As noted above, the precise form of $\Pr_i(\text{win} \mid b(v))$ depends on other agents' bidding strategies, hence the optimal strategy for agent i requires consideration of these. The standard approach in auction theory appeals to the general theory of games, and associated game-theoretic solution concepts. Specifically, we seek a *Nash equilibrium* (NE), specifying a *profile* of bidding strategies (one for each agent) such that each bids optimally given the others' designated strategies. Technically, for auction games (such as FPSB) where each agent has probabilistic beliefs about the others' valuations, the applicable solution concept is *Bayes-Nash equilibrium* (BNE).

To derive an equilibrium bidding strategy for FPSB, we make yet further assumptions. Specifically, we assume that all agents have the same valuation distribution: $F_i = F_1, i \in \{1, \ldots, N\}$. Given this symmetry of agents' positions, it makes sense to search for symmetric equilibria, that is, profiles where each agent plays the same strategy. Of course, their specific actions (bids) will be different based on varying valuations.

Note that if every agent plays the same monotone strategy, then the agent with the highest valuation will bid the most, and therefore win the good. In contemplating what strategy to adopt, then, we focus on the case where our value is highest. In order to win, we need to beat out the second-highest bidder, which by symmetry will be the bidder with second-highest valuation. Let $\bar{v}$ be a random variable that represents the highest value among the other $N - 1$ bidders. Consider the bidding policy

$$\beta(v) = \mathbb{E}[\bar{v} \mid \bar{v} < v]. \tag{3.4}$$

Under β, we bid the expectation of the highest value of all other agents, conditional on our value being the highest. This makes some intuitive sense: if we bid less than the value we expect for our strongest competitor, we provide an opening for that competitor to outbid us.

Suppose that in fact all bidders adopt β. Under symmetric value distributions, all agents have the same probability distributions for $\bar{v}$ conditional on $\bar{v} < v$. The conditional expectation for $\bar{v}$ is increasing in v, since raising the latter simply allows greater possibilities for the former. Thus, the strategy β is monotone, and the highest bidder wins. The winner's profit is $v - \beta(v)$, which in expectation is $\mathbb{E}[v - \bar{v} \mid \bar{v} < v]$, or the expected difference between the first two order statistics for the valuation distribution.

Suppose further, for example, that valuations are distributed uniformly on the unit interval, $U[0, 1]$. Then the cumulative probability distribution for $\bar{v}$ is given by $\Pr(\bar{v} < x) = x^{N-1}$, and the bidding strategy

$$\beta(v) = \mathbb{E}[\bar{v} \mid \bar{v} < v] = \frac{\int_0^v x \frac{d}{dx} \left(x^{N-1}\right) dx}{\Pr(\bar{v} < v)} = \frac{(N-1) \int_0^v x^{N-1} dx}{v^{N-1}}$$

$$= \frac{(N-1)\left(\frac{1}{N} v^N\right)}{v^{N-1}} = \frac{N-1}{N} v. \qquad (3.5)$$

In other words, each agent bids a fraction of its valuation v, depending on the number of bidders. As the number increases, we bid closer and closer to our true value. For example, if there is only one other bidder ($N = 2$) we bid half our value, and if $N = 100$ we bid 99% of our value. Intuitively, the degree of competition influences how much profit we should attempt to derive by shading.

Is bidding according to (3.5) the best we can do? Under the assumption that all other bidders adopt strategy β, the expected profit (3.3) for agent i as a function of its bid b is given by

$$\Pr_i(\text{win} \mid b)[v_i - b] = \left(\beta^{-1}(b)\right)^{N-1} [v_i - b] = \left(\frac{Nb}{N-1}\right)^{N-1} [v_i - b]. \qquad (3.6)$$

We can solve for the optimal bid by finding the extreme point of (3.6), yielding $b = \frac{(N-1)}{N} v_i$, which is in fact the bid dictated by β. What we have shown is that strategy β is optimal for the case of symmetric independent private values, with $v_i \sim U[0, 1]$, given that others have adopted this strategy. In other words, β is a symmetric BNE for this version of FPSB. It can be shown more generally [Krishna, 2010] that this strategy as given by (3.4) is a symmetric BNE for FPSB for any symmetric valuation distribution.

Going through the analysis of FPSB in such painstaking detail is worthwhile, for at least two reasons. First, it illustrates the basic machinery of auction theory, namely, analysis of the bidding situation as a game of incomplete information, culminating in verification of equilibrium bidding strategies. Second, and more directly relevant to the trading agent design enterprise, first-price trading mechanisms comprise a common auction form, especially when we include the range of variants on FPSB. Many more complex market scenarios essentially contain FPSBs within them, in the sense that they require bidding without knowledge of others' bids, and price the trade at the winner's bid. More generally, the lessons of FPSB analysis are relevant to the trading agent whenever its own bid may determine the price. Although details will vary depending on the specific context of the more complex scenario, the canonical FPSB analysis above provides a starting point.

3.2.2   SECOND-PRICE SEALED BID

A seemingly small modification of the first-price sealed bid auctions operates just as the FPSB, and likewise awards the good to the highest bidder, but assesses a price equal to the next highest bid. Stated more precisely, the *second-price sealed-bid* auction (SPSB), allocates the good to buyer $i^* = \arg\max_i b_i$, at the price $\max_{i \neq i^*} b_i$.

Despite the similarity of the auction's description, the strategic analysis of SPSB is actually much simpler than that for the first-price case. As before, we write down the agent's expected payoff as a function of its valuation and bid, assuming IPV (cf. (3.3)):

$$\Pr_i(\text{win} \mid b(v_i))[v_i - \max_{j \neq i} b_j]. \tag{3.7}$$

Note that the expression in square brackets does not depend on agent i's bid at all; if it wins, it pays the amount offered by the highest among all the *other* bidders. If this price is below v_i, then the expression in brackets is positive, and i maximizes expected payoff (3.7) by maximizing its probability of winning. Conversely, if the greatest bid by another is above v_i, then i would prefer not to win. It follows that i's optimal bid is exactly v_i. Let us consider two cases on the bid b.

1. $b \leq v_i$. If agent i wins at such a bid, the expression in brackets is necessarily nonnegative. The objective is thus to maximize probability of winning, which is achieved at the upper limit of this case, v_i.

2. $b > v_i$. A bid at v_i will already win if $v_i > \max_{j \neq i} b_j$, that is, all situations where the expression in brackets is positive. The only additional situations where i wins by bidding higher are those where the conditional payoff would be negative or zero.

Thus, we see that the optimal bid is $b(v_i) = v_i$.

Observe that our conclusion holds *regardless* of the other agents' bids. In game theory terms, this means that bidding v_i is a *dominant strategy*. From the perspective of strategy design, dominant solutions possess two highly desirable (related) features. First, the decision to adopt a dominant strategy does not depend on any assumption about other agents' behavior, beliefs, or rationality. Contrast this with the equilibrium reasoning underpinning our FPSB analysis, which is contingent on a model of other agents' valuations, their beliefs about others' valuation, and their assumption of mutual rationality. Second, implementing this strategy likewise requires no consideration of other agents. William Vickrey—whose original analysis [Vickrey, 1961] of SPSB (henceforth also known as the *Vickrey auction*) opened the field of auction theory—argued that the possibility of avoiding *counterspeculation* is a significant advantage of second-price mechanisms. As suggested by the term, any model of other agents is speculative, as their internal values and beliefs are not directly observable, nor are they unambiguously related to observable characteristics. Constructing such a model from past behavior or other features may entail significant information-gathering and cognitive cost, yet still leave substantial uncertainty. Even when the agent does have confidence in its model of others, deriving an optimal bidding strategy with respect to this model may be computationally expensive.

For all these reasons, many have followed Vickrey in advocating second-price auctions or other market designs that yield dominant strategies. Anyone wishing to simplify their bidding strategy would naturally cheer any efforts to reduce the counterspeculation required for optimal performance. They must also recognize, however, that the dominance characteristic depends on strict adherence to the SPSB analysis setting above, particularly the private value model and the assumption that this is a one-shot auction. As we see below, [4] with even small variations of or extensions to this setting, we often must consider other agents' behaviors to some degree.

It is also relevant to observe that the dominant strategy in this case is the identity function $b(v_i) = v_i$, that is, for the agent to bid its true value for the good. A mechanism that induces truthful reporting of its private information (in this case, good valuation) is called *incentive compatible* (IC), and further termed *strategyproof* if it is incentive compatible in dominant strategies. Thus, SPSB is strategyproof—there is no point to an agent strategizing, as it does best by reporting its true valuation regardless of the bidding strategies of others.

Since the bid in an IC auction corresponds to the agent's true value for the good, such a bid (and the auction) are also called *truthful*. This label is somewhat unfortunate, as it suggests that not bidding one's true value is somehow dishonest, or otherwise ethically deficient. This connotation should be resisted—truthfulness in auctions is not a moral issue. When an agent bids b in a single-item auction, it is not claiming "my value for the item is b". Rather, the semantics of such a bid is "I am willing to pay b (or less) for the item". That is, the meaning of a bid message takes the form of a commitment by the agent to engage in prospective deals, not a declaration about the agent's beliefs or preferences. The fact that the agent may value the good at more than b, or that in some circumstances it might be willing to pay more than b for it, therefore, does not bear at all on the integrity of the b bid. The agent should not incur shame for (say) shading its bid in a FPSB auction, nor should its designer harbor any guilty feelings.

Auction designers also prize IC mechanisms because identity-function strategies are conceptually simple and can render auction properties (e.g., efficient allocation) easier to analyze. A celebrated result called the *revelation principle* [Myerson, 1979] establishes that whatever outcome is achievable by a mechanism can also be achieved by some IC mechanism according to the same solution concept (e.g., dominant strategies). In that sense, IC is not really very special—it can be achieved without loss of generality, and non-IC mechanisms can produce identical results. What is special is the existence of dominant-strategy solutions, the property that allows agents to eschew counterspeculation. It is this characteristic that distinguishes SPSB, and more generally, is worth recognizing and exploiting whenever it applies in a trading environment.

3.2.3   SIDEBAR: BACK TO eBAY

The astute reader may have noted a similarity between the argument presented for truthful bidding in SPSB and the reasoning about how much to bid in an eBay snipe (Section 2.3). This is not a coincidence. In fact, the last moment of an eBay auction is essentially an SPSB mechanism. As long as the bids are submitted too late for anyone to react, any auction notifications are irrelevant, and the bidding is effectively sealed. Thanks to eBay's proxy bidding rules, the outcome is approximately that of a second-price mechanism: whomever submitted the highest (proxy) bid gets the good, at price of the second, plus an increment. This increment is why the outcomes are only approximately the same, but notice that this does not affect the optimality of truthful bidding.

---

[4]Sandholm [2000] systematically discusses a range of issues in applying the standard Vickrey analysis to computational settings.

One difference that *is* substantive is that a bid must beat the ASK quote (current high bid plus increment) to be admitted. The difference is small, because a bid high enough to be the winner but not admitted under this rule has a surplus by definition bounded by the bid increment. Nevertheless, this factor, along with tie-breaking in favor of earlier bids, provides a small countervailing incentive to bid in advance of the last moment, perhaps even taking some risk of response by incremental bidders. To a first-order approximation, although, the conclusion stands: for a one-shot eBay auction the recommended strategy is to snipe at the agent's true value for the good.

3.2.4   ENGLISH OPEN OUTCRY AUCTIONS

It might seem strange that we analyze eBay auctions based primarily on what happens at the end; such is the consequence of eBay's fixed-time ending rule. Preceding the final SPSB-like period, the auction is open and receiving bids for days to a week. As bids come in, the highest price is continually updated and displayed to bidders. This process mimics the canonical pre-Internet image of auctions, where a human auctioneer orchestrates a room of bidders, progressively raising prices until "going, going,…gone"—nobody present is willing to exceed the latest and greatest price.

This familiar auction mode, traditionally employed to sell fine art, livestock, repossessed property, and the like, is technically called an *English open outcry auction*. In the standard implementation of English open outcry, the auction proceeds in discrete rounds, each of a fixed maximum duration T_{round}. At the beginning of the round the ASK price is announced (e.g., by an auctioneer shouting "Do I hear $500?"). The first bidder to assent to this price becomes the provisional winner (designated i^*), and ASK is incremented by δ. When, in some round, the time T_{round} expires with no bidder assenting to ASK, the auction ends, and i^* is declared actual winner and pays the ASK of the previous round.

The English auction mechanism is conceptually unchanged by an asynchronous implementation, where instead of discrete rounds, the auction operates continuously over a period, accepting offers at bidders' initiative. In this version, the auction maintains a current highest bid value, BID, with $ASK = BID + \delta$. A new offer at price b is admitted iff $b \geq ASK$, and in that case the auction updates the quotes $BID \leftarrow b$ and $ASK \leftarrow b + \delta$. The auction ends at a specified closing time, with one key exception. Whenever a new bid is admitted within T_{round} of the scheduled closing time, the auction is extended to remain open for this amount of time. This provision defeats any attempt at sniping, by guaranteeing all agents sufficient time T_{round} to respond to a new bid. If at any point beyond the original closing time, T_{round} passes without any new admissible offer, the auction closes and awards the good at price BID to i^*, the submitter of the last admitted bid.

Observe that modulo reaction times and bid increments, the synchronous (discrete-round) and asynchronous versions of the English open outcry auctions are strategically isomorphic. To see this, note that any bidding strategy for the synchronous version could be directly adapted to play in the asynchronous version, and vice versa. For example, a strategy in the synchronous English auction is defined by the set A of ASK values at which the agent would bid, assuming it is not the provisional winner. In an asynchronous implementation, the agent would bid at the lowest price $\alpha \in A$ such that

$ASK \in (\alpha - \delta, \alpha]$, if such price exists and it is not already winning. Assuming that ties (multiple bidders offering in the same round or at the same time) are resolved correspondingly, an array of translated bidder strategies in the asynchronous version would lead to the same outcome as the original strategies in the synchronous English auction. A translation in the other direction would operate similarly, with equivalent results up to the bid increment (and with the same qualifications regarding tie-breaking).

Proxy bidding as in eBay (Section 2.1) essentially translates a strategy conceived in terms of a discrete-round English auction to the predominantly asynchronous mode of English auctions on the Internet. An asynchronously operating outcry auction with proxy bidding is more convenient for the bidder, who need not monitor the auction over its full duration in order to execute the intended strategy. The proxy bidding facility does not substantively alter the auction mechanism, as any strategy employing proxy bids could be implemented in a virtually equivalent manner by an agent monitoring the auction without the proxy. [5]

What is the recommended bidding strategy for English open outcry auctions? Let us adopt the independent private values model employed in our analyses of FPSB and SPSB above. Under IPV, an agent knows its own valuation at the beginning of the auction, and nothing it might observe during bidding can affect that valuation. Consider a simple strategy for the discrete-round implementation, where agent i chooses to bid in a given round iff $v_i \geq ASK$. This general approach of responding to a price quote at face value—bidding according to what the agent would wish at those prices—is called *straightforward bidding*. [6] Suppose that all other agents adopt this straightforward strategy. Then agent i can do no better than to adopt it as well. To see this, first notice that the English auction ends when ASK first exceeds what any bidder other than the provisional winner i^* is willing to pay, that is, $BID \leq \max_{j \neq i^*} v_j < BID + \delta = ASK$. The outcome is the agent with highest valuation will win the good, unless multiple agents have valuations in the final BID-ASK interval, in which case tie-breaking determines the winner among these. This winner pays BID, which is at most one increment above the second-highest valuation among the agents. If $v_i < BID$, then agent i cannot win at a profitable price, given the other-agent strategies. If $BID \leq v_i < ASK$, then agent i did bid in the last round, and won if either it was alone in doing so or selected by tie-break. Regardless, it did the right thing by refraining from the last ASK, as that would entail negative surplus. Moreover, no other strategy would allow i to win at a lower price, again given that others are following the straightforward strategy.

The reasoning above establishes that bidding in the English auction while the ASK price is below one's valuation is a Nash equilibrium strategy. Technically, it satisfies the stricter requirements for *ex post* equilibrium [Shoham and Leyton-Brown, 2009, Definition 6.3.8], since bidding this way would be an optimal response to the other agents, even if we knew all their private valuations. Thus, the straightforward approach to bidding in English auctions is a compelling strategy, assuming we

---

[5] As with many of the points in this discussion, the virtual equivalence asserted here glosses over the advantage of the proxy in reaction time due to latency in actual Internet communication. Further, proxy bidding does enable an agent to establish an earlier bid timestamp for tie-breaking purposes than any non-proxy strategy.

[6] The name is employed, for example, by Milgrom [2000] in the setting of simultaneous ascending auctions (see Section 5.4).

believe that our counterpart bidders are rational. What if they are predictably irrational, for instance as are the incremental bidders observed on eBay? For the standard English auction (in contrast to eBay), there is no clear way to exploit incremental bidders, given that the extension rule always provides them time to respond. More generally, if we knew that bidding in a certain pattern (e.g., skipping some rounds, or bidding in large increment jumps under asynchronous implementations) would change the behavior of other bidders in favorable ways, this would open up the possibility of advantage for alternative strategies. Although technically conceivable, no such strategy for English auctions has ever been shown effective and robust. [7]

For a slight variation on English auctions, termed *Japanese auctions*, the argument for straightforward bidding is more airtight. The Japanese auction is a synchronous version of English, where failing to assent to the ASK in a given discrete round eliminates the agent from the bidding henceforth. The auction ends when at most a single agent remains active. This active agent wins at the last accepted price, or if all declined, the auction randomly selects a winner among the active bidders from the previous round. The Japanese mechanism could alternatively be implemented with the interpretation that agents assent unless they explicitly decline an ASK. Under that interpretation, we could also implement a continuous-time version where the ASK ascends automatically in correspondence to a clock, stopping when the last agent but one chooses to drop out.

For Japanese auctions, the straightforward bidding strategy of dropping out when the ASK exceeds one's valuation is a *dominant* strategy. That is, this bidding rule is optimal regardless of what the other agents do. Clearly, staying active beyond the point where ASK surpasses valuation is asking for trouble. Nor can there be any advantage to dropping out before the valuation is reached: this forfeits an opportunity for a profitable trade with no countervailing benefit.

The argument for straightforward bidding in English auctions (particularly the Japanese variant) mirrors the reasoning we employed in justifying truthful bidding for the second-price sealed-bid case. Indeed, modulo the bid increment and tie-breaking, the English auction under straightforward bidding produces the same outcome as truthful bidding in SPSB. For the continuous-time clock implementation of Japanese, the correspondence is exact. In both cases, the winner is the agent with highest valuation, at a price within a bid increment (which is zero for the continuous variant) of the second-highest valuation. Although the English auction proceeds iteratively, under the private values assumption the agents do not learn anything of strategic import during the process. The analysis, therefore, collapses effectively to the one-shot case.

3.2.5 DUTCH AUCTIONS

Another common form of outcry auction is the *Dutch auction*, so called due to its prominence in Amsterdam's wholesale flower markets. The Dutch auction is a *descending-price* mechanism, that is, the ASK quote starts at a high price (above which no buyer would be willing to pay), and proceeds

---

[7]An important exception arises when participating in the bidding is costly, in which case jump bidding may deter competitors from entering by decreasing their expectations of success [Fishman, 1988, Klemperer, 2004]. Easley and Tenorio [2004] present evidence that some Internet outcry-type formats satisfy these conditions. Since automating trading strategies itself lowers bidding costs, we expect this factor to diminish in relative importance.

downward until a willing buyer steps forward to claim the good. The price is typically controlled by a clock (as in the continuous-time Japanese auction discussed above). At the moment a bidder assents to the current ASK, the auction ends, and that bidder wins the good at that price. Dutch auctions are commonly favored for their speed and simplicity of operation: important for efficient operation of open-air markets in perishable goods, like flowers or fish.

Now for the usual question: how should one bid in a Dutch auction? As it happens, we have already answered this question in another guise. A strategy for Dutch auction bidding is completely specified by the price at which the agent would claim the good, given its valuation, conditional on the auction reaching that point. The only information revealed during the auction process is that nobody is willing to bid above the current price—and even this is not an information gain since the agent could already reason that a candidate price would be relevant only under that circumstance. The bidding strategy for agent i, therefore, is a function $b_i(v_i)$. The winner of the Dutch auction is the buyer willing to come in first, $i^* = \arg\max_i b_i(v_i)$, at the price $b_{i*}(v_{i*})$. This is *exactly* the strategic definition of the first-price sealed bid auction (Section 3.2.1), thus the two auctions are *strategically equivalent*. In particular, the equilibrium bidding strategies derived under IPV for FPSB apply as well for the Dutch auction. The implications are much broader than this, however: the equilibria for FPSB and Dutch auctions necessarily coincide for *any* valuation model.

Many other auction formats have been invented and practiced over the years. [8] Some of these are small variants of the basic types discussed here, presenting similar strategic issues. Others involve more complex market situations, with multiple units, multiple goods, bidding by both buyers and sellers, or negotiation of good attributes. We address some of these complexities in subsequent sections and chapters.

3.3   INTERDEPENDENT VALUES

Much of the strategic analysis of basic auction types above hinges on the assumption of independent private values. In general, an agent's value for a good may depend on others' valuations. For example, as observed in the discussion on eBay bidding (Section 2.2), the valuation of an expert on antiques (or accordions) could sway the value of others, because it provides evidence about fundamental shared values in authenticity, craftsmanship, and other uncertain features of the good. This has direct implications for bidding strategy, in three ways. First, in an auction where information is revealed over time (e.g., the English open outcry), an agent should interpret the observed bidding behavior of others as evidence about their valuations. Second, the agent should anticipate the effect of its own bid on information revealed to others. Third, even when no information is revealed (e.g., in a sealed-bid auction), the others' bids will reflect their valuations, hence the agent should consider

---

[8]The tendency to name these after nationalities is further manifest in auction types called American, Chinese, Swedish, Yankee, and perhaps others. Whereas the Dutch auction actually does have a Holland connection, the origin of other names is somewhat more mysterious. As far as I know, one is no more likely to find a Japanese auction in Japan than anywhere else. The method of simultaneous hand signals observed in Japan by Cassady [1967] is an entirely different mechanism, and the famous Tokyo fish market currently employs Dutch auctions.

how the auction result conditional on those bids influences the probability and expected value of winning.

3.3.1   MODELING INTERDEPENDENCE

To show how these factors can be properly accounted for, we require a more precise model of the interdependence among agent values. These dependencies are relevant for bidding only when other agents' valuations could potentially inform one's own. To leave room for such influence, we must accommodate the possibility that agent i does not already know its own valuation. We do so by distinguishing what it does know—its *private information*—from its valuation. Formally, we define the agent's private information, z_i, as a *signal* that is probabilistically related to its valuation. For example, in an auction for antique furniture, an agent's signal would comprise all the information it has gathered about the condition and authenticity of the piece, as well as its background knowledge of antiques and its taste for the particular item of furniture. These are all highly relevant, but short of definitive in that the private information held by others would provide further evidence bearing on its own value for the piece.

There are several possible models of interdependent values, with qualitatively distinct structures. We can characterize the different models in terms of probabilistic graphical models called *Bayesian networks*. [9] In these models, we represent random variables by circular nodes, and dependencies among them by directed edges. For example, Figure 3.2(a) displays a Bayesian network representation of IPV. The model comprises N valuation nodes, $v_1, \ldots, v_N$, with corresponding signal nodes $z_1, \ldots, z_N$. Each agent's signal depends on its own value, and the lack of edges among the nodes for different agents reflects the absence of dependencies among their values.

We can incorporate interdependencies by positing an underlying state variable, w, which is an ingredient of every agent's valuation (in the antiques example, w might represent authenticity or craftsmanship). Figure 3.2(b) depicts this situation as a Bayesian network. This is termed a *generated signals* model [Hong and Page, 2009] because of the directed causal pathways from w to the signals. The dependence of v_i on v_j, $j \neq i$, is represented by the undirected path between these variables through w. Node w blocks this sole path, thus by the semantics of Bayesian networks, the model also entails that v_i and v_j are *conditionally independent* given w. By the same logic, the agents' signals are also marginally dependent but conditionally independent.

An extreme case of interdependent values is the *common value* model (see Figure 3.2(c)), where every agent has the same valuation, $\forall i.v_i = v$, but observes an independent signal conditional on that value. In this case, the underlying state variable w is superfluous, as the interdependence can be captured in terms of v.

The model of Figure 3.2(d) reverses the direction of arrows between values and signals compared to our generated signals model (b). In this structure, we interpret the signals as factors causing or determining the value, rather than as observed effects of these values as in the other models. For

---

[9]We cannot provide a fully self-contained treatment of Bayes nets here, but endeavor to explain the key concepts for understanding the properties of the alternative value models.

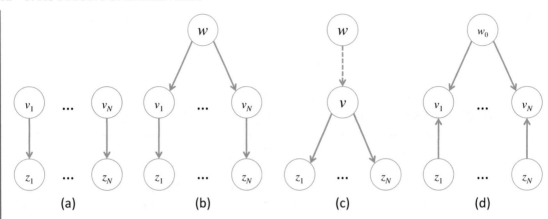

Figure 3.2: Interdependence among agents' values, expressed in Bayesian network structures. We distinguish agent i's private information (signal), z_i, from its valuation, v_i. (a) Independent private values expressed in this framework. (b) Generated signals model, where interdependencies are captured in a hidden state variable, w. (c) Common value model. (d) Signals as cause rather than observed effect of value.

example, there may be common uncertain factors (represented by w_0) relevant to all agents' value, and each i makes an observation about a feature z_i of interest only to itself. In terms of the Bayesian network, the paths between z_i remain blocked by w_0, but are now *un*blocked by the v_i, since these are *collider* nodes with both arrows pointing in. Hence, in model (d) the signals z_i are marginally independent, but conditionally dependent given the values v_i, yet still rendered conditionally independent given w_0.

These differences in dependence structure have significant implications for how agents should reason about their own value given information reflecting the other agents' values. Under IPV, we have that v_i is independent of z_j, $j \neq i$. It follows that the signal z_i is a sufficient statistic for v_i. Assuming risk neutral preferences, we can thus take $\mathbb{E}[v_i \mid z_i]$ as agent i's valuation, which explains why we do not need to distinguish values from signals in the analyses above that assume IPV.

In model (b), the agent cares about other-agent signals because they influence their bids (hence the agent's own probability of winning with its own bid), and also because they are evidence for the underlying state w. In model (d), other-agent signals provide no relevant information about the underlying state or an agent's own value. Hence, even though the values are dependent, because the signals are independent the agent can treat this for bidding strategy just like the IPV case (a).

3.3.2   WINNER'S CURSE

The relevance of interdependent values and signals becomes clear when we consider what happens if bidders ignore such relations. Let us consider a simple version of the interdependent value model (Figure 3.2(b)), as follows.

Example 3.1 Interdependent Values Suppose that w can take on two possible values, G or B, representing that the underlying value factors are "good" or "bad", respectively. Assume these cases are equally likely, and that valuations conditional on this fundamental, for all i, are given by probability density functions over $v \in [0, 1]$:

$$f_i(v \mid w = G) = 2v, \qquad\qquad (3.8)$$
$$f_i(v \mid w = B) = 2(1 - v).$$

Note that although higher valuations are more likely given G and lower values given B, these balance out and the marginal distributions are uniform, $f_i(v) = 1$ (i.e., $v_i \sim U[0, 1]$).

Consider a first-price sealed-bid auction with the value model of Example 3.1. Each agent receives a signal z_i, from which it can compute $\hat{v}_i = \mathbb{E}[v_i \mid z_i]$. From Section 3.2.1, we know that the equilibrium bidding strategy for FPSB with N bidders and independent private values distributed $U[0, 1]$ is $\beta(v) = \frac{N-1}{N}v$. Suppose $N = 2$, and agent 1 draws an uninformative observation, so that $\mathbb{E}[v_1 \mid z_1] = \mathbb{E}[v_1] = 0.5$. For simplicity, we assume the other bidder observes v_2 directly, and bids according to the FPSB IPV equilibrium strategy. This strategy calls for agent 1 to bid 0.25, and the *ex interim* [10] probability that agent 1 will win is 0.5. Under IPV, its expected profit, per (3.3), would be $0.5(0.5 - 0.25) = 0.125$. With interdependent values, however, the expectation $\mathbb{E}[v_i \mid z_i]$ cannot be used in place of v_i in (3.3). The reason is that this expectation varies, conditional on whether the agent wins the auction. In our example, agent 1 wins iff agent 2 observes $v_2 < 0.5$. This is three times more likely to occur if $w = B$ than if $w = G$, so the event where it does is evidence for worse w. In other words, winning the auction is *bad news* with respect to valuation. Specifically, in this instance $\mathbb{E}[v_1 \mid \text{win}] = \mathbb{E}[v_1 \mid v_2 < 0.5] = 0.417$. Agent 1's expected profit under this strategy is thereby reduced to $0.5(0.417 - 0.25) = 0.083$.

This *winner's curse* effect is considerably greater for larger N. In our example, if agent 1 uses the IPV policy for $N = 4$, it actually receives negative expected profit. The antidote to this curse is to anticipate the implications of the winning event on underlying state, and take this into account when bidding. Given the example $N = 4$ scenario, assuming the other three bidders use the FPSB IPV strategy, the best-response bid for agent 1 with $\mathbb{E}[v_1] = 0.5$ turns out to be 0.17, as compared to 0.375 under the IPV baseline. Shading the bid this far lifts the curse: whereas winning is still bad news with respect to valuation, agent 1 optimized the tradeoff between win probability and profit—fully accounting for this effect in advance.

3.3.3 BIDDING WITH INTERDEPENDENT VALUES

As suggested by our reasoning about winner's curse above, the key idea for dealing with interdependencies is to consider what our expected value would be, *given signals of other agents, under the*

---

[10]In discussing games of incomplete information, *ex ante* refers to the state before private information is observed; *ex interim* is the state of knowledge after each agent receives its own signal; *ex post* refers to the (imaginary) state where all private information is revealed to all.

conditions we would win. We start this time with the second-price sealed-bid auction mechanism, and proceed to cover other basic auction types.

3.3.3.1 SBSB with Interdependent Values

With interdependent values, the second-price rule does not save us from counterspeculation, as the event of winning with a given bid reveals information about others' signals, which is relevant to our own value.

As for our IPV analysis, we derive equilibrium bidding strategies for a simplified case with symmetric value distributions. Specifically, let us take the perspective of a particular bidder j, and adopt the model of Figure 3.2(b). As above, each agent receives a signal z_i, from which it can compute its private (expected) value $\hat{v}_i = \mathbb{E}[v_i \mid z_i]$. Recall that $\bar{v}_j$ denotes the maximum private value among the other bidders, $\bar{v}_j = \max_{i \neq j} \hat{v}_i$.

We consider bidding strategies expressed as a function of $\hat{v}$ (indirectly, functions of the observed signal z). A natural variant on SPSB strategy (i.e., bid true value) to account for winner's curse would be to bid what our actual expected value would be, updated to reflect the conditions under which we would win. Given symmetry, we win if no other agent has a more favorable signal. The following policy bids our expected value, conditional on the highest private value among the other agents being just as high as ours.

$$\beta(\hat{v}_j) = \mathbb{E}[v_j \mid \hat{v}_j, \bar{v}_j = \hat{v}_j]. \tag{3.9}$$

Of course, $\hat{v}$ and $\bar{v}_j$ are not generally the same. Nevertheless, suppose all agents play this strategy. By symmetry and monotonicity, agent j wins iff it is the one with the most favorable signal. If j wins, the price it pays is that of the next-highest bidder, which by assumption is $\beta(\bar{v}_j)$. As long as $\hat{v}_j \geq \bar{v}_j$ (the condition for winning), j's expected value *conditional on winning* exceeds the price it pays:

$$\mathbb{E}[v_j \mid \hat{v}_j, \bar{v}_j] \geq \mathbb{E}[v_j \mid \hat{v}_j = \bar{v}_j, \bar{v}_j] = \beta(\bar{v}_j).$$

As in our reasoning about SPSB strategy for the IPV case, bidding below (conditional) expected value risks losing the good at a profitable price, with no concomitant benefit, since the bid price does not determine the price paid, just whether the agent wins. Similarly, bidding above expected value only risks winning the good at an unprofitable price. Unlike the IPV case, however, the equilibrium strategy is *not* dominant, as the conditional expected value calculation explicitly assumes that other agents are playing the strategy β given by (3.9).

Let us examine this strategy for a particular case.

Example 3.2   (extension of Example 3.1)   Let the conditional distribution f_i on agent i's value given state w be defined by (3.8). Agents receive discrete signals $z_i \in \{L, M, H\}$, with

$$\Pr(z_i = L \mid v_i) = \frac{2(1 - v_i)}{3}, \quad \Pr(z_i = M \mid v_i) = \frac{1}{3}, \quad \Pr(z_i = H \mid v_i) = \frac{2v_i}{3}.$$

Note that the signal is probabilistically increasing in value, although the M signal is uninformative.

Suppose the situation of Example 3.2 with $N = 2$. Some relevant calculations for this scenario are presented in Table 3.1.

Table 3.1: Signal probabilities, expected values, and equilibrium bids for the scenario of Example 3.2. Agent j receives signal z_j, and the other agent receives z_{-j}. β denotes the agent's bid given these signals, and π_j denotes agent j's expected profit.

z_j	z_{-j}	$\Pr(z_j, z_{-j})$	$\mathbb{E}[v_j \mid z_j, z_{-j}]$	β_j	β_{-j}	π_j
H	H	0.123	0.700	0.700	0.700	0
H	M	0.111	0.667	0.700	0.500	0.167
H	L	0.099	0.625	0.700	0.300	0.325
M	H	0.111	0.556	0.500	0.700	0
M	M	0.111	0.500	0.500	0.500	0
M	L	0.111	0.444	0.500	0.300	0.144
L	H	0.099	0.375	0.300	0.700	0
L	M	0.111	0.333	0.300	0.500	0
L	L	0.123	0.300	0.300	0.300	0

The proposed strategy calls for agent j to bid

$$\beta(\hat{v}_j) = \mathbb{E}[v_j \mid \hat{v}_j, \bar{v}_j = \hat{v}_j] = \mathbb{E}[v_j \mid z_j, z_{-j} = z_j].$$

For example, if $z_j = M$, the agent bids according to the expected value when *both* agents receive uninformative signals. In that event, $\mathbb{E}[v_j \mid z_j = z_{-j} = M] = 0.5$, which is the same as $\hat{v}_j$, what the agent would bid under IPV. If instead, $z_j = H$, then probability calculations yield a value of $\hat{v}_j = \mathbb{E}[v_j \mid z_j = H] = 0.667$. Agent j accordingly bids the expected value given both agents observe H, that is, $\mathbb{E}[v_j \mid z_j = z_{-j} = H]$, which comes out to 0.7. Notice that interdependent values in this case lead the SPSB bidder to bid *higher* than it would in the IPV model. Intuitively, the agent reasons that the price would exceed $\hat{v}_j$ only in the case that the other agent observed a signal that increases j's expected value. For the probability distributions in this example, we also have $\mathbb{E}[v_j \mid z_j = z_{-j} = L] = 0.3$, leading the agent to bid lower than $\hat{v}_j = \mathbb{E}[v_j \mid z_j = L] = 0.333$. This can be viewed as an adjustment for winner's curse in the low signal case. In general, which cases lead to positive or negative adjustments from $\hat{v}_j$ depend on the specific distributions on values and signals.

To see that the proposed strategy constitutes a symmetric equilibrium, verify from Table 3.1 that agent j profits whenever its signal is strictly higher than the other agent's, and that no alternative bid would increase profit. For the cases where it has the same or lower signal, it receives zero (in the tie cases it is indifferent to winning), and no alternative bid would be profitable. By demonstrating that each agent's bid is a best response to the other in every situation, we establish that the strategies are actually in *ex post* equilibrium.

3.3.3.2 English Open Outcry and eBay

For an English open outcry auction, the reasoning parallels SPSB, except that observations available during the auction process may provide additional relevant conditioning information. What can be inferred depends on the precise rules of the outcry auction. At one extreme, a Japanese format auction may reveal at each round which bidders remain in the auction. In symmetric equilibrium, an agent drops out at the point where the price reaches its expected value, conditional on (1) other active bidders having equally favorable signals as itself, and (2) previously dropped out bidders having signals inferred from the information state at the point they dropped out. In principle, each agent could update its expected value on each round based on observing others' actions, and decide whether to stay in the auction. Under the assumption that all agents follow this strategy, the inferences about signals are correct and no agent could do better by bidding in some alternative manner.

Less revealing auction formats may provide only partial information about how many or which agents remain active at a given point. Nevertheless, the basic strategy idea applies: condition one's value on whatever one has observed about other agents dropping out or staying in.

To ground this analysis, let us consider the implications of interdependent values on eBay bidding strategy. Chapter 2 argues that for a one-shot eBay auction with independent values, the best strategy is to snipe at one's private value. This argument is supported by the more precise technical analysis of the IPV case in this chapter. Interdependent values add two further considerations:

1. We may observe relevant information in the behavior of other bidders prior to the auction end stage (i.e., the sniping period). In the case of eBay, we generally cannot detect bidders dropping out, however early bids do reveal information providing a lower bound on some agents' signals.

2. Submitting an early bid ourself would reveal lower-bound information on our own private value.

The second factor only underscores the motivation to snipe, as suggested in our original eBay discussion. Given that we snipe, the end of an eBay auction is essentially SPSB. The analysis above therefore applies, however our expected value calculation should also account for any information that might have been revealed by early bidders. Thus, unlike the IPV case, a sniping approach does not exempt us from monitoring the auction for relevant information, which in principle could accumulate right up to the point where the last-moment bid must be submitted.

Before leaving the eBay example, we should acknowledge a small paradox within the reasoning above. Properly accounting for interdependent values requires that we condition our expectations on information revealed by observed bids or properties of bids entailed by the event of winning the auction. These inferences necessarily invoke assumptions about bidding strategies of others, justified by rationality. However, we also recognize that some observations may be the product of irrational behavior, for example in light of our conclusion that a rational eBay bidder should always snipe. How to properly interpret observations of irrational behavior like early bidding is a deep question in game theory, and one that we do not resolve here. Rather, let us take it as a general caution against reaching overly strong conclusions based on observed behavior of others, as our models of their values and even decision processes are in practice subject to significant uncertainty.

3.3.3.3 FPSB with Interdependent Values

Bidding with interdependencies in a first-price auction follows the same basic idea to which we appealed in the second-price case: consider the expected value under the conditions we win, given assumptions about other agents' bidding strategies. Since the winning bid in FPSB sets the price, the analysis can be more complicated than the second-price case. Here we examine the scenario defined by our running example, under first-price rules. We uncover some subtle issues that did not come up in the second-price analysis.

Table 3.2 presents values, bids, and profits for the Example 3.2 environment with a FPSB mechanism, assuming that both agents employ the bidding strategy:

$$\beta(L) = 0.1; \quad \beta(M) = 0.2; \quad \beta(H) = 0.3.$$

The agent with greater signal wins the auction at its bid price. In the case of ties the winner is determined by toss of a fair coin, resulting in an expected profit of half the difference between value and price. The overall expected profit is 0.154. The strategy is *not* in equilibrium, however. Assuming the other agent plays according to β, agent j can do better by raising its bids slightly. For example, if it bids 0.301 on the H signal, it now wins that case outright regardless of the other-agent signal, paying only slightly more. The revised strategy yields an expected profit of 0.178.

Table 3.2: Scenario of Example 3.2, $N = 2$, for a FPSB auction.

z_j	z_{-j}	$\Pr(z_j, z_{-j})$	$\mathbb{E}[v_j \mid z_j, z_{-j}]$	β_j	β_{-j}	π_j
H	H	0.123	0.700	0.3	0.3	0.200
H	M	0.111	0.667	0.3	0.2	0.367
H	L	0.099	0.625	0.3	0.1	0.325
M	H	0.111	0.556	0.2	0.3	0
M	M	0.111	0.500	0.2	0.2	0.150
M	L	0.111	0.444	0.2	0.1	0.244
L	H	0.099	0.375	0.1	0.3	0
L	M	0.111	0.333	0.1	0.2	0
L	L	0.123	0.300	0.1	0.1	0.100

This example suggests a difficulty in finding equilibrium strategies when the signals are discrete but the bid prices range over a continuum. In a first-price auction, the agent seeks to win by as small a margin as necessary, which with continuous prices is not well defined. Ties are likewise unstable, as the agent can break out of the tie situation by an arbitrarily small increase in bid. This issue may be averted if the signals are continuous and generated by an atomless distribution (i.e., any point value has infinitesimal probability), since ties are essentially impossible and any finite change in price corresponds to a change in probability of winning. Recall the equilibrium strategy for FPSB exhibited in Section 3.2.1. The problem did not arise there because the observation (exact valuation rather than a signal in the IPV case) distribution is continuous.

We can also sidestep the problem if our action space is discrete. For example, suppose bids must be submitted in multiples of 0.1. Given that restriction, the strategy examined in Table 3.2 is in fact a best response to itself.

3.4   STRATEGY LESSONS

Let us summarize some of the key strategic insights from basic auction theory presented in this chapter.

1. In a first-price auction, the price is determined by the winner's bid. The bidder should therefore *shade* its bid in order to ensure positive surplus when it wins.

2. In a second-price auction, an agent's bid determines whether it wins, but does not affect the price conditional on winning. It is therefore optimal to bid one's true value, regardless of the other agents' bids.

3. Open outcry auctions operate quite differently than sealed-bid mechanisms, but the strategic considerations are parallel. Dutch auctions are strategically equivalent to FPSB, and English auctions correspond quite closely to SPSB.

4. When agents' values are interdependent, the bidding strategy should take into account all evidence gleaned about other-agent valuations during the course of the auction, as well as any information that adheres to the conditions under which a given bid will be successful. In particular, when agent values are positively related, the bid should be adjusted in consideration of the winner's curse—learning that one's signal is the most favorable is "bad news" with respect to value.

More complex auctions go beyond the basic types analyzed in this chapter. Nevertheless, insights from the basic auctions often carry over, for example implications of the winner's bid setting the price or not (i.e., the general distinction between first- and second-price mechanisms), and cautions about winner's curse in the face of interdependent values. Strategies from the basic auctions may be directly relevant when considering auction mechanisms that are composites of the basic auctions. For example, as we discussed for eBay shopping, the strategy for dealing with a one-shot market can be modified to account for the repeated opportunities that often present in real trading environments.

Sequential composition is one way that complex auctions are founded on basic auction building blocks. Another is simultaneous auctions, the focus of Chapter 5. Auctions may also be more complex in that they operate iteratively, providing dynamic feedback to bidders as the auction proceeds over time. This dynamism is a key characteristic of the continuous double auction we investigate in the next chapter. As we find in these examinations, even a little bit of dynamics or simultaneity tends to get us past where auction theory can provide definitive answers. Despite the lack of neat solutions, there is ample basis for strategic guidance even in these complex market situations.

3.5   BIBLIOGRAPHIC NOTES

Cassady [1967] provides an informative overview of the history of auctions, from ancient origins to mid-20th Century practice. Lucking-Reiley [2000] chronicles the use of second-price auctions as far back as 1893.

William Vickrey's pioneering study [Vickrey, 1961] of the four classic auction types (FPSB, SPSB, Dutch, English) provided a frame for the subsequent development of auction theory. Research over the ensuing half century broadened the range of auction mechanisms analyzed, and deepened our understanding of strategic bidding issues, particularly with respect to the structure of incomplete information and its revelation through a bidding process. Numerous literature surveys and textbook treatments of auction theory are available [Klemperer, 2004, Krishna, 2010, McAfee and McMillan, 1987, Milgrom, 1989, 2003]. These sources provide an entry to the topic of auction design, as well as risk aversion, collusion, budget constraints, and many other relevant issues ignored or glossed here. The text of Shoham and Leyton-Brown [2009] includes game theory background and a chapter on auction theory written primarily for computer scientists.

The analysis of auctions under IPV is standard since Vickrey, and covered by all the references above. Issues related to asynchronous operation, proxy bidding, and other salient features of Internet versions of English open outcry are discussed by Steiglitz [2007]. The Japanese auction was first modeled formally (and termed a *button auction*) by Milgrom and Weber [1982].

The phenomenon of winner's curse was first identified by [Capen et al., 1971], based on the authors' experiences in bidding on oil drilling leases. Their article is also noteworthy as an interesting early use of computer simulation to gain insight into bidding strategies. Thaler [1988] discusses empirical evidence for winner's curse, both in practice and in laboratory studies.

Milgrom and Weber [1982] introduced a general framework for interdependent values and qualitative characterization of the correlations among signals and values. The same paper also derived equilibrium bidding strategies for these models. Our discussion of signal dependence structures and their implications for reasoning about interdependent values (Section 3.3.1) was inspired by Hong and Page [2009].

CHAPTER 4

Continuous Double Auctions

One of the most basic trading scenarios is an abstract market based on the *continuous double auction* (CDA) mechanism [Friedman, 1993]. The CDA is a simple and well-studied auction institution, employed commonly in commodity and financial markets. Indeed, all of the major stock exchanges, as well as commodity and futures trading institutions, employ some form of CDA for the bulk of their trading activity. This was true 100 years ago before digital computers existed, and remains the case in today's era of electronic trading environments.

The "double" in CDA refers to the fact that both buyers and sellers submit bids. CDA markets are "continuous" in the sense that they mandate trades instantaneously on receipt of compatible bids.

4.1 CDA OPERATION

The rules for a basic CDA market are quite simple. A bid comprises an offer to buy or sell various quantities of a good at corresponding prices. For example, an agent might offer to buy 10 units at a price of $50 each. A more complex bid might include multiple offers, for example to buy 10 units at $50 and another 10 (i.e., 20 in total) if the price is $40 or less. We refer to the compound offer as a single bid (in order to maintain our convention of at most one active bid per agent per auction), although in actual operation the market may track them separately.

In financial markets, a bid specifying price is termed a *limit order*, in contrast to *market orders* which offer to buy or sell specified quantities at any available price. If the market is highly *liquid*, that is, one can depend on the existence of many competitive standing offers to buy or sell, submitting a market order will usually result in a transaction at a price near the current quote. Nevertheless, there is generally little reason to avoid specifying a limit price just in case,[1] and so we restrict attention to bids as limit orders for the remainder of this discussion.

In a continuous auction, offers transact as soon as they are matched. It follows that at any given time, the set of active bids in a CDA, called the *order book*, contains no matches. We generally refer (for CDAs as well as any double auction) to the match operation as *clearing* the market, since any compatible offers are cleared out of the order book.

---

[1]Some markets or brokerages charge reduced fees for market versus limit orders. Retail traders should be wary of such discounts, however. Whereas brokers are obliged to seek best available executions under US security regulations, the operational definition of these requirements leaves some wiggle room, and other factors such as payments for order flow may affect routing of market orders. In any case, the risk of price movement in the time it takes a retail order to be filled weighs heavily compared to small fee discounts.

As shown in Figure 4.1, we can visualize the CDA order book as composed of two lists, one each for buy and sell offers, with offers sorted by price.[2] Let the buy offers be sorted such that the highest price is first, and the sell offers sorted starting with lowest price. With this organization, we can provide a price quote simply by reading the first element of each list. The BID quote is the price of the highest buy offer, and the ASK that of the lowest sell. The difference between these, called the BID-ASK *spread*, is the range of prices for which no trades are possible with incumbent offers.

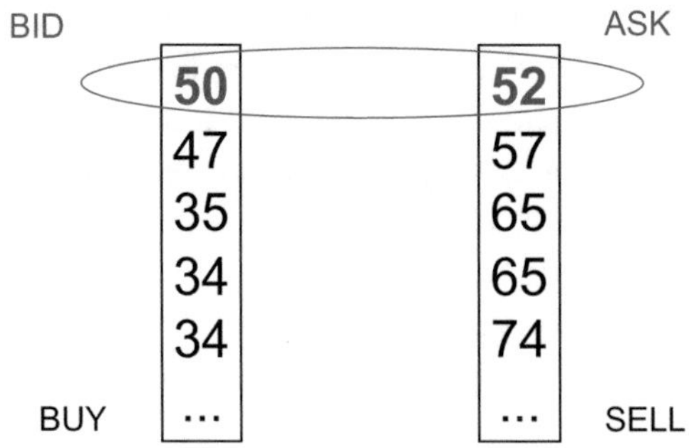

Figure 4.1: CDA order book, organized into sorted lists for buy and sell offers. Each entry shown here is a limit order for one unit. Multiunit offers would be annotated with an explicit quantity specification.

When a new bid is received, we can quickly determine whether there is a match by comparing with the BID and ASK prices. If there is a match, then the new offer trades with the agent who submitted the most attractive matching incumbent offer, at the price of that incumbent offer. The incumbent offer (or part of it that matched with the new bid) is then removed from the order book. If there is no match, the new bid is simply added to the order book, and sorted buy/sell lists maintained as appropriate. There may also be a partial match, where some offers in the bid (or some units in the case of multiunit offers) are matched but not all. In this case, the matched portion trades as dictated above, and the remaining part is added to the order book. For example, if a new offer to buy four units at $60 were submitted to the order book in Figure 4.1, the result would be a transaction of one unit at $52, one at $57, and insertion of two units at $60 at the top of the buy list.

The CDA is a limiting special case of a more general class of double auctions called *call markets*. In a call market, the auction clears at designated times, rather than continuously. To clear the market, the auction determines a single price at which the number of units offered to buy equals that offered to sell (for offers exactly at this price, the auction may include them or not). Between clears, offers are accumulated in the order book. At any time, the ASK price (not necessarily revealed as a quote to bidders) is defined as the minimum price at which a new offer to buy a single unit would be successful, given the rest of the order book. Similarly, BID is the maximum price at which a new single-unit sell offer would be successful. The auction's clearing policy dictates how the price is determined within the range of BID and ASK.

---

[2]Wurman et al. [1998] provide a detailed discussion of data structures and algorithms for double-auction order books.

For example, in the order book shown in Figure 4.2(a), for any price $77 \leq p \leq 85$, there are two units offered for sale, and two for purchase. This range of feasible clearing prices thus defines the *BID-ASK* spread. In the order book with multiunit offers of Figure 4.2(b), the unique price balancing supply and demand is 67. The three offers to buy a total of 20 units exceeding that price would transact with 20 units of the sell order right at that price.

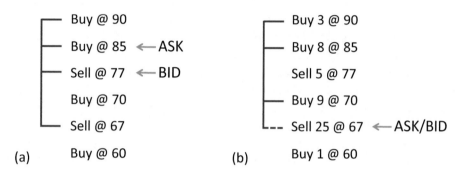

Figure 4.2: Call market order books: (a) single-unit offers, (b) multiunit offers. Connecting lines indicate offers that would transact if cleared in the current state.

This example and the discussion above more generally assumes that offers are *divisible*, that is, that they may be traded in whole or partial quantity, at the associated unit price. Some double auctions allow specification of *all-or-none* (AON) bids, which require that they be matched in full in order to execute a trade. An AON bid that matches only in part is added to the order book, just as if it did not match. Price quotes are no longer straightforwardly interpretable as thresholds to transact a unit, given the possibility of such constraints. Many markets that admit AON bids process new bids for potential matches in a greedy manner [Miller, 2002], although more sophisticated optimization approaches may generally be preferred [Schvartzman and Wellman, 2007].

4.2   ONE-SHOT DOUBLE AUCTIONS

Before considering the strategic problem of CDA bidding, it is helpful to analyze a simpler variety of double auction, where bids are accumulated without price quotes (i.e., in a sealed manner), and the auction clears exactly once, at a specified time. The *one-shot double auction* is a call market with a single clear event. We demonstrate that the bidders' strategy for a one-shot DA depends on the auctions' clearing policy, and on whether the agent seeks to trade more than one unit.

4.2.1   SINGLE-UNIT TRADERS

We start with the case where each trader is interested in buying or selling only a single unit of the good. Let the number of sell offers in the order book at a given time be M. If we sort all of the offers (both buy and sell) by price, as in Figure 4.2, then the Mth highest price corresponds to ASK, and the $(M + 1)$st highest to BID. To see this more generally, suppose there are n buy offers among the M highest, by definition at the ASK price or better. (For a CDA, $n = 0$ at all times by construction, and for the call market order book of Figure 4.2(a), $M = n = 2$.) This leaves $M - n$ sell bids among this group, and therefore $M - (M - n) = n$ sell bids at the BID price or lower. Thus, we can clear these n highest buys and n lowest sells at any price $BID \leq p \leq ASK$, resulting in a state where all unmatched sell bids have a price of ASK or greater, and all unmatched buys are at BID or less. This is therefore consistent with the clearing criterion and our definition of price quotes.

Suppose our rule is to clear at the $(M + 1)$st (BID) price. A buyer in this case optimizes its expected surplus by bidding exactly its (independent private) value, by the same reasoning we employed for the SPSB auction (Section 3.2.2). [3] The buyer wins the good iff its offered price is among the top M bid prices, and conditional on being in that group its actual bid price has no influence on the price it pays—the $(M + 1)$st. Thus, the $(M + 1)$st-price one-shot DA is strategyproof (incentive compatible in dominant strategies) for buyers. On the other hand, the seller does not generally optimize by reporting its true value, since it may win with an offer that in fact determines the $(M + 1)$st price.

By a symmetric argument, the Mth-price one-shot DA is strategyproof for single-unit sellers, but not buyers.

It is also possible to set a clearing price somewhere between the Mth (BID) and $(M + 1)$st (ASK) prices. The *k-double auction* [Satterthwaite and Williams, 1993] is a one-shot DA, where the interpolation parameter $k \in [0, 1]$ defines the clearing price:

$$(1 - k)BID + kASK.$$

Thus, the 0-DA clears at the BID price, and the 1-DA at the ASK. For $0 < k < 1$, the auction clears strictly between the two (unless $BID = ASK$), and the auction is not incentive compatible for either buyers or sellers. Given probability distributions for buyer and seller valuations, the setup can be analyzed as a Bayesian game. In equilibrium, buyers will shade their bids, offering somewhat below their true value, and sellers will shade theirs upward. Satterthwaite and Williams [1993] analyze the k-DA in depth, deriving bounds on the deviation of equilibrium bids from true value, and on the deviation of the auction result from perfect efficiency. As we would expect, the degree of loss due to strategic bidding decreases with the number of bidders.

4.2.2   MULTIUNIT TRADERS

For the case of multiunit offers, we compute price quotes as above, with M the number of units offered for sale. For example, in the order book of Figure 4.2(b), $M = 30$, and both the BID and ASK quotes coincide with the offer to sell at 67. The number of units offered to buy at that price or greater is $n = 20$, and so that is the total quantity that would be traded at market clear.

---

[3]SPSB corresponds to the special case where $M = 1$ and the seller bids zero. The reasoning extends straightforwardly to positive seller bids (reserve prices).

When agents are interested in trading multiple units, we lose incentive compatibility even for the boundary cases of the 0-DA and 1-DA. Consider a buyer with value for multiple units. Under the 0-DA, the price b_j the buyer offers for unit j cannot affect the price it pays *if it wins j units*. However, the price it pays for j units *may* be determined by its offer for the $(j + 1)$st unit: for instance in the case j of its offers are in the top M, and its $(j + 1)$st offer constitutes the $(M + 1)$st highest price overall.

More concretely, consider an agent who would like to obtain six units of the good and values them at 75 each. Suppose the market adopts the $(M + 1)$st-price clearing rule, and the other-agent bids are as listed in Figure 4.2(b). If the agent submits an offer Buy 6 @ 75 to this order book, the market will clear at a price of 70 (the 31st-highest price once we insert this agent's six units), and thus the agent will achieve a surplus of 30 (six units at a surplus of five each). Suppose instead the agent submits the offer Buy 5 @ 75. In this situation the clearing price is 67, and the agent obtains five units at eight surplus each, for a total of 40.

What this example shows is that the agent is better off not offering to buy the sixth unit, even though it is available at a price, 70, below its value, 75. Whereas this unit is profitable on its own, from this agent's perspective it also imposes a cost of three on acquiring each of the first five units. These first five are called *inframarginal*, as they are obtainable at prices strictly within the margin of the threshold price of the sixth unit. The total cost of the sixth unit is effectively $70 + (3 \times 5) = 85$, which exceeds its value. [4]

The strategy of attenuating bid quantity in order to obtain better prices is called *demand reduction* [Ausubel and Cramton, 2002]. Demand reduction provides a way for an agent to account for the effect of its own bids on prices. The basic idea is to bid for each unit at the price at which its cost—inclusive of impact on inframarginal units—equals the unit's value. Given a model of the price impact of one's own bid, one could in principle derive an optimal bid for any particular situation.

4.3   CDA STRATEGIES

4.3.1   STRATEGIC ISSUES FOR CDAS

All of the strategic issues for one-shot DAs apply as well to continuous DAs. Specifically, to the extent an agent's bid may determine the price it pays or receives, the agent has an incentive to shade the offer away from its true value. Recall that according to CDA rules (Section 4.1), the price of each trade is determined by the incumbent offer. Thus, a bid determines its potential clearing price iff it is not matched immediately. Even if it is matched immediately, the transaction modifies the order book (annihilating offers at the top of the other side), potentially impacting the price of future trades. Multiunit traders therefore face a strategic imperative to reduce demand, just as in the one-shot case.

CDAs also present strategic issues arising from the dynamic nature of the auction mechanism. In dynamic auctions, the agent must consider *when* to bid as well as how much, or more generally, how

---

[4]The suboptimality of truthful bidding is not merely an artifact of the $(M + 1)$st-price clearing rule. In fact, there does not exist any way of setting a uniform DA clearing price to achieve incentive compatibility for buyers or sellers with multiunit values [Wurman et al., 1998].

much as a function of time. This time element—completely absent from one-shot auctions—bears on strategy in three key ways. First, the agent may have a time preference over trading outcomes, that is, it may care about when it trades. Second, the agent must anticipate the temporal pattern of prices, as determined by other bidders' behavior. Third, the agent's own bids get reflected in price quotes, which reveals information to the other bidders.

In general, auction theory has not successfully tackled these dynamic strategy issues in a setting approaching the richness of CDAs. The most advanced effort along these lines was a model developed by Wilson [1987], in which a set of agents interested in trading a single unit interact through a CDA mechanism. The results are theoretically illuminating, [5] but do not capture observed phenomena and have not been realized in a computationally operational bidding strategy. Most subsequent research on CDA strategy has adopted an experimental approach, either based on human subjects or simulation with software bidding agents.

4.3.2   EXPERIMENTAL CDA RESEARCH

The CDA has been widely employed in experimental economic studies, and notably in an open research competition conducted at a Santa Fe Institute workshop in 1990 [Friedman and Rust, 1993, Rust et al., 1994]. The winning trader in this competition—named Kaplan after its author—held back until most of the other agents revealed their valuations through bidding behavior, then "stole the deal" by sniping at an advantageous price. Agents employing more elaborate reasoning failed to make such sophistication pay off. This is consistent with observations that even extremely naive strategies achieve virtually efficient outcomes in this environment. The most striking demonstration along these lines employed agents who bid randomly according to a uniform probability distribution: what Gode and Sunder [1993] dubbed the "zero intelligence" (ZI) strategy. As long as the bids were constrained not to produce a loss (i.e., buy bids were bounded above by value, and sell bids bounded below by cost), the authors found that experimental markets with ZI agents converged to equilibrium prices and achieved close to maximal global surplus. Although such results suggested limits on the potential global returns to positive smarts, it does not preclude the possibility of individual agents doing much better on their own account.

Over the last 20 years, CDA markets have provided the context for many further studies of artificial trading agents. The simplicity and familiarity of the abstract CDA framework presents some distinct advantages as the basis for trading agent research. These include ease of explanation and simulation, low barriers to entry, consensus understanding of market rules, predictability of behavior, opportunity to build on prior work (on design of both mechanism and agents), and analyzability of outcomes. Given the ubiquity of the CDA institution, there is also a potential to incorporate real-world market data of various kinds.

---

[5]Wilson constructs a possible equilibrium where traders use willingness to delay (and risk not trading before time expires) as a way to signal value. After waiting enough time to establish it is the highest-valued buyer (or lowest-valued seller), an agent decides to bid, and trades with its counterpart. The counterparties always act simultaneously and correctly in this equilibrium, so there is never a non-null order book.

4.3.3 ADAPTIVE CDA BIDDING STRATEGIES

One important family of bidding strategies works by adapting its bid relative to underlying valuation, based on observed market conditions. The thread of research on these strategies had its origin in zero intelligence. Cliff [1997] observed that ZI agents may not generally converge to equilibrium prices, particularly if the demand and supply are not symmetric. He proposed a new strategy, called "zero intelligence plus" (ZIP), to address the limitations of ZI. Although the name suggests an incremental extension, ZIP actually improves individual performance dramatically over zero intelligence.

The ZIP strategy uses a simple form of machine learning, adjusting requested profit margins based on market conditions. Agents begin with an initial target price, τ, set at a random deviation from their valuation for the good (below for buyers, above for sellers). The agent continually adjusts its bid price toward the target. At each observed bid, the agent perturbs τ up or down based on how the bid compares to its own current price, and whether the bid was successful. The adjustment method is a form of Widrow-Hoff delta rule [Russell and Norvig, 2009, Chapter 21], and employs several tunable parameters including a learning rate and a momentum term.

Since its introduction, ZIP has been widely tested, enhanced, and refined. Tesauro and Das [2001] extended ZIP to employ an array of target profit margins (one per traded unit) instead of a single value. Preist [1999] introduced a simpler update rule. Cliff [2009] ultimately produced a version of ZIP that included 60 parameters tuned with a genetic algorithm.

Vytelingum et al. [2004] proposed a "risk-based" (RB) strategy that, much like ZIP, gradually adjusts expected margins based on market conditions. The strategy employs a parameter $r \in [-1, 1]$, termed the *aggressiveness factor*, which governs the relation between the agent's attempted profit margin and historical prices. Let $v_{\max}$ denote the maximum possible valuation for the good, and p^* a moving average of historic trade prices. A seller calculates its target price τ by

$$
\tau = \begin{cases} p^* + (v_{\max} - p^*)re^{(r-1)\theta} & \text{if } 0 \leq r \leq 1 \\ p^* + (p^* - v_i)re^{(r+1)\left[\log\left(\frac{v_{\max}-p^*}{p^*-v_i}\right)-\theta\right]} & \text{if } -1 \leq r < 0, \end{cases} \tag{4.1}
$$

where $\theta \in [-1, \infty)$ is a fixed parameter that specifies the rate of change in τ with respect to r. An example is shown in Figure 4.3. Note that sellers with $r > 0$ demand prices above prevailing levels, whereas those with $r < 0$ seek lower but surer profits.

Sellers use a delta rule to increase r when trade occurs at a price $p \geq \tau$, and decrease it when trading at $p < \tau$ or submitting sell bids at $p \leq \tau$. On each bidding opportunity, the actual bid price is set to $p = ASK - (ASK - \tau)/\eta$, with η another parameter. Buyers apply a symmetric approach.

Further development of RB produced a refined strategy called "adaptive aggressiveness" (AA) [Vytelingum et al., 2008]. AA extends RB by using a delta rule to adapt θ to market conditions, increasing (decreasing) the rate of change of τ given a higher (lower) measure of price volatility α. Let $(\theta_{\min}, \theta_{\max})$ and $(\alpha_{\min}, \alpha_{\max})$ define ranges over which the variables θ and α, respectively, are updated. Target values of θ are given by

$$
\theta_{\min} + (\theta_{\max} - \theta_{\min}) \left(1 - \bar{\alpha}e^{\beta(\bar{\alpha}-1)}\right),
$$

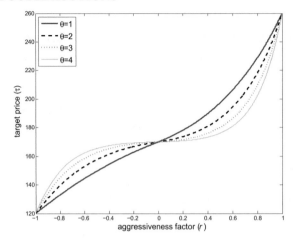

Figure 4.3: Target price as a function of r and θ, as computed by sellers using the RB strategy, with $p^* = 170$, $v_{\max} = 260$, and $v_i = 120$.

where $\bar{\alpha} = \frac{\alpha - \alpha_{\min}}{\alpha_{\max} - \alpha_{\min}}$, and β is a fixed parameter that determines the rate of adjustment in θ. AA's target price computation is similar in spirit to that of RB (4.1), with some specific differences as described by Vytelingum et al. [2008].

4.3.4   HEURISTIC BELIEF LEARNING

Another particularly influential trading strategy was proposed by Gjerstad and Dickhaut [1998], later revised and termed the "heuristic belief learning" (HBL) model [Gjerstad, 2007]. An HBL agent maintains a belief state over acceptance of hypothetical buy or sell offers, constructed from historical observed frequencies. It then constructs optimal offers with respect to these beliefs and its underlying valuations. The timing of bid generation is stochastic, controlled by a *pace* parameter, which may depend on absolute time and the agent's current position. Gjerstad [2007] demonstrates that pace is a pivotal strategic variable, and that indeed there is surprisingly large potential advantage to strategic dynamic behavior despite the eventual convergence to competitive prices and allocations.

More specifically, the HBL strategy employs observations during a CDA game instance to construct a *belief function* representing the probability of a bid being accepted depending on its price. For a seller, the belief function is given by

$$\Pr(p) = \frac{TS(p) + B(p)}{TS(p) + B(p) + US(p)}, \tag{4.2}$$

where $TS(p)$ is the total number of transacted sell bids at a price p or higher, $B(p)$ is the total number of buy bids submitted to the auction at a price p or higher, and $US(p)$ is the total number of unmatched sell bids up to p. The function value is calculated for every price p included in the history of bids (submitted by all agents) within a specified window, and extended to the positive reals using cubic spline interpolation. The seller bids at a price p that maximizes expected surplus, defined as $\arg\max_{p>v} \Pr(p)[p-v]$, given a cost of v. Buyers apply a symmetric heuristic.

Note that HBL presumes that the agent observes all bids submitted to the auction. According to the CDA rules as presented in Section 4.1, an agent actually can detect only changes to price quotes, so offers that are worse than current quotes enter the order book unobserved. Experimental CDA studies that include HBL typically accommodate the strategy in one of two ways (neither especially realistic for CDA environments): extending observability to all bids, or restricting bids to beat the current price quote.

The original HBL was further refined by several authors, for example, MGD [Tesauro and Das, 2001] [6] modifies the belief function to reduce volatility. In extensive simulated trials, the authors found that MGD outperformed a range of other strategies, including ZI, ZIP, and Kaplan—the sniping strategy that won the original Santa Fe tournament. The strategy also compared favorably with human traders [Das et al., 2001].

HBL attempts to optimize surplus, but in a myopic fashion, considering only the next immediate trade. To address this limitation, Tesauro and Bredin [2002] proposed a further extension, dubbed GDX, that optimizes bids over time using dynamic programming (DP). GDX's DP formulation represents a state by the agent's pending trades and remaining bidding opportunities, and estimates transition probabilities using the HBL belief function (4.2).

The approach maximizes discounted cumulative future rewards rather than just immediate profits. Specifically, the expected value of a state, given i trades left and t remaining bidding opportunities, can be expressed by a Bellman equation:

$$V(i,t) = \arg\max_p \big(\Pr(p)[\sigma_i(p) + \gamma V(i-1, t-1)] + [1 - \Pr(p)]\gamma V(i, t-1) \big),$$

where $\Pr(p)$ is given by (4.2) and $\sigma_i(p)$ is the surplus obtained by trading the next unit at price p. These expected values are stored in a table, as part of the DP algorithm. GDX weights rewards using a discount parameter γ, such that as $\gamma \to 0$, GDX reproduces the basic HBL strategy, whereas with $\gamma \to 1$ GDX differs the most from HBL.

In order to determine a new bidding price, GDX recalculates its DP table on every bidding iteration. Tesauro and Bredin [2002] found experimentally that high values of γ create a strong GDX trader that clearly outperforms basic HBL in a wide variety of market scenarios.

4.3.5   STRATEGIES FROM FINANCE RESEARCH

Because CDAs or close variants are widely employed in financial markets, models from the finance literature that account for details of the trading mechanism, or *market microstructure* [Garman, 1976, O'Hara, 1995], are also highly relevant to trading agent strategy. Much of this literature addresses the trading problem from the perspective of a *market maker*: an agent who posts limit

[6] Most of the literature continues to label the HBL strategy "GD" after the original authors, hence the "modified GD" (MGD) strategy name for the variant. We employ the name HBL for the original strategy here because it is more descriptive, and also the apparent preference of the lead author, Gjerstad.

orders on both buy and sell sides, aiming to profit on the bid-ask spread by trading across incidental fluctuations [Chakraborty and Kearns, 2011]. Market-maker strategies may be based on simple heuristics, inventory control techniques [O'Hara, 1995], or machine learning methods [Das, 2005].

Availability of real-time market information has recently begun to enable higher-fidelity modeling of financial trading environments. The Penn Exchange Simulator [Kearns and Ortiz, 2003] merges bids from automated trading agents with actual limit-order streams, providing realistic volume and volatility patterns, whether or not these would emerge naturally from the artificial agent strategies. Competitions based on this simulator enabled comparison of a wide variety of CDA bidding policies [Sherstov and Stone, 2004], including some that may use information from the entire order book [Kearns and Ortiz, 2003].

4.4   EMPIRICAL GAME-THEORETIC ANALYSIS

Researchers continue to search through the strategy space in quest of better CDA bidding strategies. Many of these studies support their conclusions with outcomes from particular configurations of strategies (e.g., uniform mixtures of the strategies compared), self-play, round-robin tournaments, or some combination of these. Given that the performance of a strategy depends on those played by other agents, however, results can be highly sensitive to this choice of strategic context. A demonstration of a strategy's effectiveness is compelling in proportion to the plausibility of its assumptions about other-agent behavior. This is of course why game theory relies so centrally on equilibrium concepts for strategic reasoning. Although the complex market games induced by CDA mechanisms resist conventional game-theoretic analysis, we can adapt experimental techniques to incorporate equilibrium-based principles. We have been developing such a methodology in our own research over several years, and call this approach *empirical game-theoretic analysis* (EGTA).

4.4.1   EGTA METHODOLOGY

The basic EGTA process is illustrated in Figure 4.4. The first step is to define a restricted strategy set. For most trading games, the full strategy space is enormous: many dimensions, with large or even uncountable cardinality. In the CDA, for example, the strategy dictates what order(s) to place at each time, as a function of private information (value or cost for each unit of the good), as well as all observed history (price quotes and transactions). Exploiting environmental regularities (e.g., symmetry, independence) renders game-theoretic analysis possible for a small set of basic auction models (see Chapter 3), but has generally not been sufficient to tackle dynamic markets like the CDA. Instead, we identify a manageable number of strategy candidates, and analyze the game restricted to those strategies. Choice of candidates may be based on simplifying assumptions, rules of thumb, experience, or high-level agent design principles.[7] For example, the CDA strategies discussed above were introduced by researchers with ideas about particular strategic issues. Our inclusion of these in the EGTA study described below inherits these motivations, reinforced by the literature's reported

---

[7]Walsh et al. [2002] refer to candidates in a restricted game as *heuristic strategies*. As the phrase suggests, the strategies have a basis in problem-solving knowledge, but we generally lack a theoretical characterization of the ideal set.

experience with these strategies. Since many of the strategy descriptions include tunable parameters (e.g., γ for GDX), a candidate set built this way is naturally extensible across several well-defined dimensions.

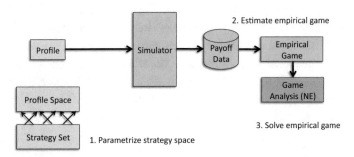

Figure 4.4: Basic steps of empirical game-theoretic analysis (EGTA).

The set of strategy candidates induces a space of profiles (assignments of strategies to agents) for the game. To evaluate strategies in a profile, we typically have no more direct method than to simulate the strategies in the given market environment. To account for probabilistic influences (distribution of private information and potentially other stochastic factors), we generate many samples of each profile, collecting for each sample the payoff (realized utility) of each strategy. From the accumulated samples, we induce a game model, termed the *empirical game*. This model may be a straightforward estimate (e.g., using sample means for each evaluated profile), or based on a more complex statistical approach [Jordan and Wellman, 2009, Vorobeychik et al., 2007].

Given the game model, we may analyze it using any of the standard tools of game theory, such as computing Nash equilibria.

4.4.2   EMPIRICAL CDA GAME

A few experimental investigations of CDA strategy have incorporated some form of equilibrium analysis, in order to address the issue of strategic context raised above. Phelps et al. [2006] and Vytelingum et al. [2006] appealed to *evolutionary* stability to characterize solutions to restricted CDA games. Walsh et al. [2002] conducted an EGTA study of three CDA strategies—ZIP, HBL, and Kaplan. Their empirical game has multiple equilibria, including one where all three strategies are played with positive probability. The case for HBL is particularly compelling, however, as they observe that this strategy would become nearly dominant with only a 5% improvement.

In the most comprehensive study to date, Schvartzman and Wellman [2009] evaluate using EGTA all the prominent strategies proposed in prior experimental CDA literature. Specifically, we implemented a generic CDA environment as similar as possible to those employed in previous works. Using the empirical game-theoretic analysis approach described above, we simulated all combinations of seven strategies: Kaplan, ZI, ZIbtq (ZI where bid prices are constrained to *beat t*he current price *q*uote), ZIP, MGD (the variation of HBL introduced by Tesauro and Das [2001]), GDX, and RB. For a four-player CDA game [8] with seven strategies, there are 210 distinct profiles, each of which we sampled over 100 times to generate estimated payoffs. The result constitutes an empirical game model over the restricted strategy set.

Analyzing this game produces strong support for HBL-based approaches to CDA bidding. Specifically, MGD and GDX were the only survivors of iterated elimination of dominated strategies. Everyone playing GDX is a Nash equilibrium of this game, as is the profile with three playing MGD and one GDX. We also identified an approximate NE, where players choose MGD with probability 0.68, and GDX otherwise. Table 4.1 ranks the seven strategies by *NE regret*, the amount lost by one player when deviating to a given strategy from a Nash equilibrium. The ranking here is for regret with respect to "NE1", the profile in which all players choose GDX.

Table 4.1: Average NE regret, symmetric profile regret, and symmetric profile payoff of each strategy in the empirical CDA game. NE1 is the all-GDX equilibrium of the game with strategies from the literature. NE2 is the equilibrium comprised of new RL-derived strategies.

strategy	NE1 regret	NE2 regret	symm. regret	symm. payoff
GDX	0	1.32	0	247.98
MGD	0.49	3.26	1.11	248.57
RB	2.20	8.64	3.76	248.08
ZIP	2.90	9.86	5.72	247.95
Kaplan	4.56	24.55	367.83	2.02
ZIbtq	14.67	17.44	40.08	247.45
ZI	16.42	16.82	48.36	248.07

The table also lists the regret (potential benefit from deviation) as well as payoff for profiles where everyone plays the corresponding strategy. The simple sniping strategy (Kaplan) is relatively viable when other agents play a sophisticated strategy like GDX, but disastrous when everyone adopts it. Whereas all the rest perform reasonably well in self-play, the ZI-based profiles are likewise quite unstable.

The analysis confirms the general superiority of HBL variants observed in prior literature. Employing reinforcement learning (RL) techniques [Szepesvári, 2010], we were able to derive new strategies that outperform these. Specifically, we defined an environment where one player generated behavior using RL, and the rest played according to the current empirical equilibrium. In our RL formulation, the action (bid) is expressed in terms of offset from marginal value (i.e., the immediate surplus demanded), and state in terms of a collection of relevant features, including statistics summarizing price history, current price quotes, units to trade, and time. We applied this approach iteratively: learning a strategy that deviates from the current equilibrium (i.e., provides greater profits

---

[8]In this analysis, each player chooses the bidding strategy for four agents—two buyers and two sellers, for a total of 16 agents in the simulation. This reduction produces an approximation of the full 16-player game, at a small fraction of the computational cost [Wellman et al., 2005].

when played in the equilibrium context), calculating a new equilibrium incorporating the learned strategy, and repeating until no improvement is found.

This interleaved EGTA/RL process succeeded in producing effective new strategies for the generic CDA environment. These learned strategies represent the current champions for this environment, in the sense that in equilibrium, one would choose learned strategies exclusively among the set comprised of these plus those from prior literature.

This improvement was achieved despite the fact that HBL and other strategies use significant history and bid information not assumed available in our RL formulation. Moreover, 135 parametric variations of MGD, GDX, and AA all failed to break into the equilibrium of learned strategies.

Unfortunately, the RL-derived strategies have no simple description, and thus cannot be easily transferred to other environments. The interleaved EGTA/RL method used to derive them, however, is generally applicable. For instance, the same methods used to generate new champion strategies for the generic CDA also proved successful in finding improved strategies for the CDA trading component of the TAC travel-shopping game [Schvartzman and Wellman, 2010]. Nevmyvaka et al. [2006] also demonstrated the promise of RL techniques for CDA strategy in a large-scale study based on real financial trading data.

4.5   LESSONS FOR CDA STRATEGY

The continuous double auction is a ubiquitous trading mechanism, employed widely to trade financial securities, commodities, and virtually any kind of standardized good. Although simple to describe, the CDA defies analytic solution—no equilibrium bidding strategy is known, nor any general technique to derive optimal responses to plausible models of other traders. Nevertheless, designers of trading strategies for CDAs can draw on several principles and strategy ideas from prior investigations of this domain. Here we summarize some of the key lessons presented in this chapter.

1. Employ adaptive techniques to improve a baseline strategy. ZIP and its successors demonstrate that starting from arbitrary targets and adjusting based on experience can produce results competitive with approaches that more explicitly model the environment.

2. Construct predictive models from observations. The HBL family of strategies uses price quotes and bid observations to assess the probability of winning across a range of bid levels.

3. Account for the impact of one's own bids on the market. Attenuating quantity through *demand reduction* can enhance profits for units traded, more than offsetting the nominal profits of units foregone.

4. Look ahead to anticipate future market states. GDX employs dynamic programming to account for the effect of near-term actions on subsequent trading opportunities.

5. Evaluate strategy candidates through systematic simulation modeling. Typically, there is no more direct way to assess the outcome of a given strategy than to simulate its behavior in a

plausible environment model. In the EGTA approach, simulations cover sufficient strategy combinations (profiles) to reason in terms of game-theoretic solution concepts.

6. Employ reinforcement learning or related methods to generate strategies for new environments. Using parametric versions of known strategies, or structured representations of bidding policies, one can specify a well-defined space of strategy candidates. In conjunction with simulation or other model forms, it is possible to automate the search over this space for the most compelling strategy profiles.

It is important to acknowledge some assumptions underlying these prescriptions. Effective adaptation or modeling from observations requires that the agent has access to sufficient information at execution time to learn its environment. Building simulation models, in contrast, requires that the designer can produce a plausible model of the trading environment (including its uncertain elements) in advance. Accounting for own price effects, anticipating future market trajectories, or otherwise exploiting predictive information in turn depends on either having an advance model or being able to learn quickly enough from available experience. Given the inherent unreliability of advance modeling or fast learning, some additional effort to evaluate the robustness of strategies to environmental assumptions is always recommended.

CHAPTER 5

Interdependent Markets

One of the most attractive features of automated trading is the ability to monitor and participate in many markets simultaneously. Compared to human traders, a software agent can take in data from multiple sources at high throughput, and in principle process a massive quantity of information relevant to trading decisions in short time spans. As described in Section 1.3, an index arbitrage strategy monitors the relation of an index security to the underlying basket of securities defining the index, and triggers a trade whenever a sufficient disparity is observed. Such strategies are necessarily automated, for by the time a human noticed the arbitrage opportunity it would be gone. Indeed, as noted in the Introduction, "program trading" is sometimes defined (e.g., by the New York Stock Exchange), in terms of strategies that simultaneously transact bundles of goods.

Dealing with multiple markets also poses one of the greatest strategic challenges for automated trading. When markets interact, a strategy for bidding in one must consider the implications of and ramifications for outcomes in others. We tasted a flavor of such issues in preceding chapters, for instance: shopping across multiple eBay auctions (Chapter 2), and trading multiple units in CDAs (Chapter 4). In this chapter, we take on the problem of multiple interdependent markets[1] directly, framing the fundamental questions and introducing some general concepts useful for analyzing cross-market interactions. In the process, we expand our repertoire of generic auction models to include simultaneous markets, both one-shot and iterative versions.

5.1 DEALING WITH MULTIPLE MARKETS

Markets are *interdependent* when an agent's preference for the outcome of one market depends on the outcomes in other markets. We model this by extending the value function to apply over *bundles* of goods. Let $\mathcal{X}$ be the set of goods allocated by the collection of markets. Then $v_i : 2^{\mathcal{X}} \to \mathbb{R}$ represents agent i's value for the subsets of items it may acquire in the aggregate market outcome. Due to interdependence, an agent's value for an individual good is relative to the other goods it obtains. We express such relative individual-good preferences in terms of *marginal value* (MV).

Definition 5.1 Marginal Value Agent i's *marginal value*, $\mu_i(x, X)$, for good x with respect to a (fixed) bundle of other goods X is given by:

$$\mu_i(x, X) = v_i(X \cup \{x\}) - v_i(X), \tag{5.1}$$

---

[1]N.b. the concern here with interactions across markets, in contrast with the issue of interdependent *values* (Section 3.3), which refers to interactions across *agents*.

where $X \cup \{x\}$ is the bundle obtained by adding the good x to X.

If the bundle of goods (second argument of μ) matters, we have interdependence.

Definition 5.2   Market Dependence   The market for allocating good x depends on the other markets iff there exist j, X, and X', $x \notin X \cup X'$, such that $\mu_j(x, X) \neq \mu_j(x, X')$.

The challenge in dealing with multiple interdependent markets lies in two implications of the dependence for decision making:

1. *Exposure.* Decisions in one market must be taken under uncertainty, before the outcomes of all relevant other markets are realized. In particular, placing a bid exposes the agent to the risk that it may obtain a good at a price it would not have wanted to pay had it known the results of other markets.

2. *Influence.* Decisions in one market may affect outcomes in other markets, through the behavior of other agents.

Note that the second implication may apply even to an agent i with preferences not exhibiting interdependence. For instance, the fact that agent i wins good x may influence how j bids for good y, thus affecting i's outcome in the latter market. These influences can be somewhat subtle, so we start by focusing on the exposure challenge posed by interdependent markets.

Example 5.3   Market-based scheduling   Suppose three agents are competing to obtain reservations for a meeting room, which are allocated in auctions for one-hour slots. Agent 1 has a three-hour meeting, starting at time 1. Agents 2 and 3 have one-hour meetings, and they prefer earlier times to later. The table below displays the bundle value functions for this scenario.

Name	$v(\{1\})$	$v(\{2\})$	$v(\{3\})$	$v(\{1, 2\})$	$v(\{2, 3\})$	$v(\{1, 3\})$	$v(\{1, 2, 3\})$
Agent 1	0	0	0	0	0	0	15
Agent 2	8	6	5	8	6	8	8
Agent 3	10	8	6	10	8	10	10

If the goods (the three reservation slots) are allocated in separate markets, the uncertainty about joint outcome presents difficulty for the agents' bidding decisions—regardless of the particular auction rules. Suppose the auctions proceed sequentially. Then agents 2 and 3 have a reasonable option, namely, the "shopping" strategy discussed in the context of eBay auctions (Section 2.4). For each successive good, this strategy would bid based on marginal value, discounted to reflect future market opportunities. But for agent 1, this strategy clearly would not work, as marginal value is zero unless it already has two slots. For this agent to have any chance of surplus, it must obtain all three goods. It must therefore bid on the first item, even though it does not know whether it will be able to win the other two at an acceptable price.

Agent 1's quandary is an instance of the *exposure problem*: to obtain a valuable bundle it must risk getting stuck with a set of goods it would not have wanted at the resulting prices. In this case, the risk is that it will obtain a strict subset of the desired goods. The problem arises in this form because the agent's preference for the goods is *complementary*, which means that it obtains more value for some goods conditional on also obtaining some others.

Definition 5.4 Complementary Preferences   A value function exhibits *complementary preferences* if there exist bundles of goods X and Y such that

$$v(X \cup Y) > v(X) + v(Y).$$

In agent 1's case, the complementarity is extreme: the reservation slots are worthless unless it can obtain all three. If subsets had some value (i.e., a shorter reservation would be acceptable although less preferred), the exposure issue could still present in a less drastic form. The difficulty is precisely attributed to *increasing marginal value*.

Definition 5.5 Increasing Marginal Value   Agent i's value function exhibits *increasing marginal value* for x iff there exist y and X such that:

$$\mu_i(x, X \cup \{y\}) > \mu_i(x, X).$$

In Example 5.3, agent 1 faces increasing marginal value: for instance, $\mu_1(1, \{2, 3\}) = 15 > 0 = \mu_1(1, \{2\})$. A value function that does not exhibit increasing MV is said to obey *diminishing marginal value*. Marginal value is diminishing, intuitively, when having more goods can only make an additional good less valuable.

Whenever there is increasing MV, the agent potentially faces the exposure risk of failing to complete its desired bundle, as it must bid for x before it knows whether it can obtain y. In our scheduling example, the agent's strategy for bidding in the first auction would have to trade off the benefit of getting the full reservation with the risk of falling short, based on its best information about the prospects of subsequent auctions. The class of value functions exhibiting increasing marginal value contains those exhibiting complementarity.

The exposure problem is only more difficult when the markets operate simultaneously rather than sequentially. With serial markets the agents can at least condition their decisions on outcomes from prior auctions. For simultaneous markets, even agents 2 and 3 face an exposure problem, although their preferences have no complements. In their case, the problem is that they may bid on more than one slot given the uncertainty of outcome for each, and risk obtaining extra (thus worthless) goods at positive price.

The suboptimality of market outcomes due to exposure and other challenges of interdependent markets can to some degree be alleviated by coordinating market allocation across goods. *Combinatorial auctions* [Cramton et al., 2005] tackle the problem directly, by soliciting bids for entire bundles, and deriving optimal allocations based on these. Although such mechanisms may provide an effective solution in many cases, the need to coordinate across goods often presents significant barriers to adoption [MacKie-Mason and Wellman, 2006]. Even when some combinatorial allocation is feasible, its scope is inevitably limited. For example, the goods an agent cares about may be distributed across markets that are operated by different parties. Indeed, to some extent virtually all goods will be connected by through some dependencies for some set of agents, and clearly it would not be practical to allocate all the world's goods through a single overarching mechanism. Hence, there is ultimately no avoiding the fact of multiple markets for related goods.

5.2   SIMULTANEOUS ONE-SHOT MARKETS

We start with a simple model of simultaneous markets, relaxing restrictive assumptions in a series of steps. Consider m auctions, each for one item, $\mathcal{X} = \{x_1, \ldots, x_m\}$. To begin, suppose the agent knows the prices at which it could obtain the respective goods. For example, this would be the case if the auctions were *posted-price markets*, that is, where the seller specifies a price and the buyer decides whether to take it or leave it. [2] The *known-price assumption* removes all uncertainty, and so the agent faces no exposure problem. Although this case does not reflect the real challenge of interdependent markets, the known-price solution can serve as a starting point for constructing bidding strategies for less idealized scenarios.

Given a fixed vector of prices, $\boldsymbol{p} = \langle p_1, \ldots, p_m \rangle$, let $\sigma_i(X, \boldsymbol{p})$ denote agent i's surplus from obtaining the set of goods X at those prices:

$$\sigma_i(X, \boldsymbol{p}) \equiv v_i(X) - \sum_{j \mid x_j \in X} p_j. \tag{5.2}$$

The agent faces a straightforward problem of optimizing surplus.

Definition 5.6   Acquisition Problem   Given prices $\boldsymbol{p}$, the *acquisition problem* [3] selects optimal goods to purchase:

$$X^* = \mathrm{ACQ}_i(\boldsymbol{p}) \equiv \arg\max_{X \subseteq \mathcal{X}} \sigma_i(X, \boldsymbol{p}),$$

where σ is defined by (5.2).

An optimal strategy for the known-price case would be to compute a solution $X^* = \mathrm{ACQ}(\boldsymbol{p})$ and buy the goods X^*. Its result is the optimal surplus at these prices:

$$\sigma_i^*(\boldsymbol{p}) \equiv \sigma_i(\mathrm{ACQ}_i(\boldsymbol{p}), \boldsymbol{p}).$$

---

[2] Absolute certainty about prices also entails that the seller has no supply constraints, and so is guaranteed to fulfill all orders at these prices. Such a fixed-price mechanism may not seem worthy of the term *auction*, but recall our inclusive definition (Section 3.1).
[3] The generalized version of ACQ allows for initial holdings and multiunit auctions as well [Boyan and Greenwald, 2001]. The *completion problem* extends acquisition to allow selling as well as buying goods.

5.2.1 BIDDING WITH GIVEN PRICES

Let us next consider a scenario where the agent has to submit bids rather than simply announce its chosen bundle. For concreteness, we take the auctions to operate according to second-price (SPSB) rules. Each auction grants its good to the highest bidder, at a price equal to the next highest, which could include the seller's reserve price. The known-price assumption in this context means that the agent has a correct estimate of what it would have to bid to win, and the price it would pay if successful.

Given known prices p, the optimal bidding strategy for simultaneous SPSB auctions is straightforward. The agent computes an optimal bundle $X^* = \text{ACQ}(p)$, as above, and submits bids $b_j > p_j$, for all $x_j \in X^*$. Under our assumptions, the agent indeed obtains X^*, and the prices ratify the optimal result by construction. Since the amount bid does not affect actual prices paid under second-price rules, the outcome holds regardless of the bid amounts as long as they exceed the given price. For concreteness, consider the strategy of bidding at just above the given prices for goods in the target set. We accordingly call this strategy TargetPrice. Let $\epsilon > 0$ be an arbitrarily small price increment.

$$\text{TargetPrice}_j = \begin{cases} p_j + \epsilon & \text{if } x_j \in X^*, \\ 0 & \text{otherwise.} \end{cases}$$

Unfortunately, TargetPrice is highly sensitive to the assumptions on which it is based. If prices are only slightly higher than anticipated, the agent fails to get X^*, even if it would have been willing to pay more for the goods lost. We would prefer strategies that retain optimality under these assumptions, but may exhibit more robustness to their violation. For this we appeal to the concept of marginal value, extended to account for price information.

Let $p[p_j \leftarrow q]$ be a version of the price vector with the jth element revised as indicated: $p[p_j \leftarrow q] = \langle p_1, \ldots, p_{j-1}, q, p_{j+1}, \ldots, p_m \rangle$.

Definition 5.7 Marginal Value with Buying Opportunities   Agent i's *marginal value* $\mu_i(x_j, p)$ for good x_j with respect to prices p is given by:

$$\mu_i(x_j, p) = \sigma_i^*(p[p_j \leftarrow 0]) - \sigma_i^*(p[p_j \leftarrow \infty]). \tag{5.3}$$

Here, $\sigma_i^*(p[p_j \leftarrow 0])$ represents agent i's optimal surplus (solving its acquisition problem) at the given prices, assuming it receives good x_j for free. Similarly, $\sigma_i^*(p[p_j \leftarrow \infty])$ represents the optimal surplus if x_j were unavailable. The difference is precisely the marginal value of good x_j with respect to its buying opportunities for other goods. Note that Definition 5.7 can be viewed as a generalization of Definition 5.1, under the interpretation that goods in X have zero price, and all other goods have infinite price.

Consider Example 5.3 with $p = \langle 4, 4, 4 \rangle$. Agent 1's marginal value for good 1 given the opportunity to buy at prices p is:

$$\mu_1(1, \langle 4, 4, 4 \rangle) = \sigma_1^*(\langle 0, 4, 4 \rangle) - \sigma_1^*(\langle \infty, 4, 4 \rangle) = 7 - 0 = 7.$$

Agent 3's marginal value for this good is $\mu_3(1, \langle 4, 4, 4 \rangle) = 10 - 4 = 6$.

Marginal values can provide a valid guide to bidding in some circumstances. The principal exception is under conditions of increasing marginal value, which as we illustrated above can present difficult exposure issues. The following property rules this out.

Definition 5.8 Diminishing Marginal Value at Prices Agent i's value function exhibits *diminishing marginal values* at prices p iff for all j and j':

$$\mu_i(x_j, p[p_{j'} \leftarrow 0]) \leq \mu_i(x_j, p).$$

The relativity of this definition to prices indicates that MV might be increasing at some prices, and diminishing at others. In particular, we may have diminishing MV without buying opportunities (i.e., the complement of Definition 5.5 holds), but not at some particular prices.

The diminishing MV property dictates that the marginal value for good j does not increase when some other good j' becomes available for free. We could have equivalently expressed the condition in terms of *any* decrease in $p_{j'}$. The only way a price change for any good can affect MV for another is through a change in optimal acquisition. If j' is in the optimal acquisition already, decreasing its price has no effect. Otherwise, there is some threshold less than $p_{j'}$ (possibly as low as zero) at or below which it is acquired, and above which it is not. Therefore, the condition $\mu_i(x_j, p[p_{j'} \leftarrow 0]) \leq \mu_i(x_j, p)$ will hold iff $\mu_i(x_j, p[p_{j'} \leftarrow p_{j'} - \epsilon]) \leq \mu_i(x_j, p)$ holds for all $\epsilon > 0$.

The reliability of MV under these conditions is established by the following proposition. Let X^* be a solution to the acquisition problem at prices p: $X^* = \text{ACQ}(p)$. Given diminishing marginal value at prices p, it can be shown that [Wellman et al., 2007]:

- $\mu(x_j, p) \geq p_j$ if $x_j \in X^*$, and

- $\mu(x_j, p) \leq p_j$ if $x_j \notin X^*$.

Marginal value and price coincide, $\mu(x_j, p) = p_j$, iff there are multiple solutions to ACQ and x_j is in some but not all.

Let StraightMV be the strategy of bidding marginal value for all goods,

$$\text{StraightMV}_j = \mu(x_j, p).$$

If the solution to ACQ is unique, and the prices are correct, then by the proposition above the agent will win exactly the goods in the optimal acquisition. If there are multiple solutions to ACQ, however, the StraightMV agent will acquire the union of goods in optimal acquisitions, which is generally not itself optimal. To avoid this problem, we introduce a revised strategy, TargetMV, which selects a single solution X^* to ACQ, and bids marginal value only for those goods.

$$\text{TargetMV}_j = \begin{cases} \mu(x_j, \boldsymbol{p}) & \text{if } x_j \in X^*, \\ 0 & \text{otherwise.} \end{cases}$$

This strategy is guaranteed to produce an optimal result, regardless of the uniqueness of ACQ solutions, as long as the assumptions of diminishing marginal value and price correctness hold.

Since the TargetMV strategy refrains from bidding on goods outside X^*, arguably they should be irrelevant to bid calculations. Strategy TargetMV* is just like TargetMV except that it calculates marginal values treating goods not in the acquisition set as if they are unavailable (priced at ∞).

$$\text{TargetMV*}_j = \begin{cases} \mu(x_j, \boldsymbol{p}[p_\ell \leftarrow \infty \mid x_\ell \notin X^*]) & \text{if } x_j \in X^*, \\ 0 & \text{otherwise.} \end{cases}$$

Without the opportunity for alternatives, the marginal values of goods in X^*, and therefore the bids, will be higher.

Under the known-price and diminishing marginal value assumptions, all of the targeted bidding strategies (TargetPrice, TargetMV, and TargetMV*) are optimal, and StraightMV is optimal as well if the solution to acquisition is unique. Which is the better strategy, therefore, hinges on how well they perform when the assumptions are violated. To illustrate, suppose the market-based scheduling instance (Example 5.3), with all agents predicting prices $\boldsymbol{p} = \langle 6, 5, 2 \rangle$. Table 5.1 shows how each agent would bid in this instance for each of the strategies introduced in this section. At the specified prices, agent 3's acquisition problem has two solutions—{1} and {3}—and we assume it chooses the former in its targeted strategies.

Table 5.1: Agent bid vectors for the market-based scheduling scenario, $\boldsymbol{p} = \langle 6, 5, 2 \rangle$, under targeted and marginal-value strategies.

	TargetPrice	StraightMV	TargetMV	TargetMV*
Agent 1	$\langle 6, 5, 2 \rangle$	$\langle 8, 7, 4 \rangle$	$\langle 8, 7, 4 \rangle$	$\langle 8, 7, 4 \rangle$
Agent 2	$\langle 0, 0, 2 \rangle$	$\langle 5, 3, 3 \rangle$	$\langle 0, 0, 3 \rangle$	$\langle 0, 0, 5 \rangle$
Agent 3	$\langle 6, 0, 0 \rangle$	$\langle 6, 4, 2 \rangle$	$\langle 6, 0, 0 \rangle$	$\langle 10, 0, 0 \rangle$

Although agent 1's valuation exhibits increasing marginal value (Definition 5.5), in this instance the marginal values for goods in the target bundle still exceed the prices. Thus, as long as the prices are correct, and ties are broken favorably, any of the strategies produces the desired result (winning all three goods) for agent 1. Similarly, agent 2 wins good 3 (its optimal choice) at these prices. Agent 3 wins good 1 under all strategies, however with StraightMV it also wins good 3—a profitable but suboptimal outcome attributable to the multiplicity of acquisition solutions.

However, the assumption of accurate pricing (and favorable tie breaking for each agent) is not actually consistent with this set of bids. For example, if all agents bid using StraightMV, agent 1 wins all three goods at prices $\langle 6, 4, 3 \rangle$. This is a good outcome for agent 1, but not strategically stable. For instance, if both agents 1 and 2 play StraightMV, agent 3 would be better off playing TargetMV*, thus winning good 1 (at price 8), and leaving agent 1 to win the other two goods (worthless without the first) at a price of 3 each. We can continue this line of reasoning, in effect solving a game over the set of heuristic strategies considered, as in empirical game-theoretic analysis (Section 4.4). In this case, we would find that TargetMV* is a dominant strategy for agents 2 and 3 (best for any choices of the other agents), and that when the latter agents play TargetMV* it does not matter which strategy agent 1 chooses (it wins good 2 at zero price regardless). These strategic conclusions are particular to this problem instance, and would not generally hold for different configurations of agent valuations or (incorrect) price predictions.

Nevertheless, we can make one general comparison. Given diminishing marginal value, strategy TargetMV* is always at least as good as TargetMV, no matter how wrong the price prediction, and regardless of the bidding strategies of other agents. Both strategies bid for the same set of goods, and given that choice of target, the actual prices for other goods is irrelevant. In either strategy, the prices bid for the target goods are at least what was predicted, so if these turn out lower nothing changes. The only differences that matter are for prices greater than predicted, which could mean losing the corresponding good under TargetMV or both cases. TargetMV* bids the marginal value for each good with respect to buying opportunities for the target set. By diminishing marginal value, we know that if an element of the target set is lost, this can only increase the marginal value of others. Thus, the agent still wishes to buy the goods at prices up to (and possibly beyond) what it bids under TargetMV*, even if the prices for other goods turn out to be incorrect. Since TargetMV* prices are weakly greater than TargetMV prices, the former strategy necessarily performs at least as well as the latter.

5.2.2   BIDDING WITH PRICE DISTRIBUTIONS

In actuality, a bidder faces substantial uncertainty about the prices at which it could obtain desired goods. For instance, in our running example, agent 1 might believe that the prices will be either $\langle 6, 5, 2 \rangle$ or $\langle 8, 7, 4 \rangle$. At the lower prices it would prefer to obtain the bundle, but at the higher prices would not. In this case, a bid of $\langle 6, 5, 2 \rangle$ (or slightly higher) would ensure it gets the goods iff the prices are favorable, given these price possibilities. Suppose instead it believes the prices will be $\langle 6, 5, 2 \rangle$ or $\langle 16, 5, 2 \rangle$. In this case, there is no bid vector that obtains the bundle in the favorable case, without risking negative profits in the alternative. Whether it should bid at all depends on the relative probabilities of the two outcomes.

To account for uncertainty, let us relax the known-price assumption and consider bidding strategies that take as input a probability distribution F over prices. We retain the property (by virtue of employing second-price auctions) that the agents' bids do not determine the price for goods they win. The agent's optimal strategy is to submit a vector of bids that maximizes expected surplus with respect to the price distribution. As above, we seek heuristic strategies that determine bids in terms of sensible concepts like marginal value.

One straightforward approach is to approximate the distribution by its mean-value prices, and apply known-price bidding strategies such as those introduced in the preceding section. That is, once we calculate the average prices (by Monte Carlo sampling or any other method), we can employ these as input to TargetPrice, StraightMV, TargetMV, or TargetMV*. This is an application of the standard *expected value method* for approximating the solution to a stochastic optimization problem [Birge and Louveaux, 1997].

An alternative approach is to use the distribution of prices to calculate a distribution of marginal values, and determine bids based on these. In particular, we might sample from the price distribution, and for each sample calculate the marginal values at those prices. We could then take the sample mean of these marginal values, and treat this as a marginal value as in the MV-based strategies already discussed. For instance, let AverageMV denote the strategy that calculates average marginal values, and simply bids these values.

$$\text{AverageMV}_j = \mathbb{E}_F[\mu(x_j, \boldsymbol{p})].$$

Note the distinction: StraightMV given a price distribution bids the marginal values at average prices, whereas AverageMV bids the average marginal value with respect to the distribution of prices.

The contrast is illustrated in Table 5.2, which displays the bid vectors produced for the strategies discussed in this section, for our running example. The calculations are based on predicted prices distributed independently and uniformly for the respective goods: $U[0, 12]$, $U[0, 10]$, $U[0, 4]$. Note that the mean price vector for this distribution is $\langle 6, 5, 2 \rangle$, so the strategies based on the expected value method produce the same bids presented in Table 5.1.

Table 5.2: Agent bid vectors for the market-based scheduling scenario, under distribution-based strategies, assuming predicted prices $\langle U[0, 12], U[0, 10], U[0, 4] \rangle$.

	TargetPrice	StraightMV	TargetMV	TargetMV*	AverageMV
Agent 1	$\langle 6, 5, 2 \rangle$	$\langle 8, 7, 4 \rangle$	$\langle 8, 7, 4 \rangle$	$\langle 8, 7, 4 \rangle$	$\langle 8, 7.00, 4.47 \rangle$
Agent 2	$\langle 0, 0, 2 \rangle$	$\langle 5, 3, 3 \rangle$	$\langle 0, 0, 3 \rangle$	$\langle 0, 0, 5 \rangle$	$\langle 4.49, 2.07, 1.85 \rangle$
Agent 3	$\langle 6, 0, 0 \rangle$	$\langle 6, 4, 2 \rangle$	$\langle 6, 0, 0 \rangle$	$\langle 10, 0, 0 \rangle$	$\langle 5.14, 2.61, 1.40 \rangle$

These strategies are heuristic approximations, and which is best is contingent on the particular problem instance. For agent 1, it turns out that all of these heuristic strategies yield negative profits at the predicted price distribution, thus it is better to not bid at all and obtain zero. Agent 2 does best with strategy AverageMV (expected profit 3.46), whereas agent 3's top strategy is StraightMV (4.7) in this instance. Thus, even with the same predicted prices, the agents' valuations dictate different heuristic strategies.

Although none of these simple heuristics is adequate as a universal bidding strategy, we may combine them to yield a more powerful bidding method. Since evaluating a candidate bid vector with respect to a price distribution is relatively easy, we can use simple heuristics (these and others) to generate candidates, and then select the best for the actual bid. Let us call this the BidEvaluator approach, denoting a family of strategies parametrized by the methods for generating candidates to evaluate. In addition to the strategies considered, we can also generate candidates by sampling from the price distribution and applying the various known-price strategies to each sample.

Let us consider BidEvaluator(TargetMV), a version of BidEvaluator that samples from the price distribution and generates candidates by applying TargetMV to the sample. For agent 1, this strategy produces a bid to guarantee winning all three goods (assuming the price distribution), for an expected profit of 2 (the optimum). For agents 2 and 3, it produces bids that equal the performance of TargetMV* (expected profits of 3 and 4.17, respectively). More expansive methods for generating candidates may approach the optimal bids for these agents. For this problem, an agent 2 bid of $\langle 5, 2, 2 \rangle$, is approximately optimal (expected profit slightly better than AverageMV). Agent 3 achieves a profit of 4.81 with bid $\langle 5.5, 2.5, 2 \rangle$.

BidEvaluator is just one of several possible approaches to bidding optimally with respect to price distributions. One method shown effective in bidding problems is *sample average approximation* [Greenwald et al., 2009, Kleywegt et al., 2002], which optimizes bids with respect to a sample from the price distribution.

Finally, note that as in the previous section, when prices are determined endogenously, they do not generally confirm the beliefs of the agents. In this case, when agents bid optimally with respect to the input distribution, a second-price auction would result in agent 1 getting all the goods, at prices $\langle 5.5, 2.5, 2 \rangle$. This is consistent with the given distribution, but just one possible realization. Moreover, the resultant prices are not stable—agents 2 and 3 would both demand goods at those prices. We address the issue of aligning price beliefs with results directly below, in our discussion of methods for price prediction.

5.3   PRICE PREDICTION

Predictions about prices, whether in the form of point price vectors or probability distributions, play a pivotal role in strategies for bidding in interdependent markets. Such predictions provide a way to deal with the uncertainty inherently posed by interdependence. Contrast the situation for isolated second-price auctions, where the uncertainty about other-agent behavior is irrelevant (assuming IPV), since the bidder has a dominant strategy. With interdependent markets, even with IPV and second-price auctions, the other agents are relevant because the value of the good in each auction may depend on outcomes of other auctions. The most general approach would thus express uncertainty over other-agent bids, or more fundamentally, over their valuations and strategy choices. Given the complexity of modeling other agents' individual actions, however, we seek a more aggregate model of market state. Prices provide a convenient summary of the agent's opportunities across markets, capturing in a single number (per good) the net market influence of other agents.

To this point we have taken price predictions as given inputs to the bidding problem. It bears asking, where do these predictions come from? The general source is experience from interacting with or observing similar markets. Estimates may come directly from aggregating past data, or through more complex mapping of observed historical patterns to features of the state of the market at hand. Experience may also be augmented by models, including both theoretical economic pricing models as well as structural models induced from historical data.

Analysis of bidding strategies developed by participants in the annual Trading Agent Competition (TAC) over the years provides insight about potential sources and methods for price prediction. TAC events present researchers with challenging market game scenarios, involving interdependent markets and many other strategically difficult features. An investigation of strategies developed for the TAC Travel game revealed that (1) a significant majority of entrants included explicit price-prediction modules, and (2) a diverse set of techniques based on summarization of historical data, machine learning, and theoretical modeling all enjoyed some success [Wellman et al., 2004]. Particularly effective were machine learning approaches that accounted for instance-specific features of the observed market state [Stone et al., 2003]. Purely theoretical models based on competitive equilibrium analysis (discussed below) proved comparably accurate in this domain [Cheng et al., 2005].

Further indication of the variety of prediction methods comes from the TAC Supply Chain Management game [Pardoe and Stone, 2010a]. In this context, agents must bid in interdependent markets to buy input components and sell finished products, and inevitably successful strategies entailed serious attention to the price prediction subproblem. Agents employed a variety of machine learning methods, including linear regression [Benisch et al., 2006], nearest-neighbor [Kiekintveld et al., 2009], particle filtering [Pardoe and Stone, 2009], and combinations. Given the diversity of market environments and plethora of plausibly predictive features, it is not surprising that the range of methods for price prediction would span the toolbox of machine learning. That similar methods developed in one market context can be applied to another (e.g., Pardoe and Stone report adapting the particle-filtering approach to a qualitatively different scenario in bidding for ad auctions [Pardoe and Stone, 2010b]) is evidence for the genericity of price-prediction problems.

A survey of potentially relevant prediction techniques would necessarily cover much of machine learning, and thus is beyond the scope of this volume. In lieu of such broad treatment, we discuss a few model-based methods that are specific to the economic and auction context.

Competitive analysis [Arrow and Hahn, 1971] models markets under the assumption that agents take prices as given (i.e., perfect competition), and markets equilibrate. In our terms, a *competitive equilibrium* (also called *Walrasian equilibrium*) is a price vector $\boldsymbol{p}^*$ such that the aggregation of good bundles solving the agents' acquisition problems matches the goods available (not counting goods at zero price). That is, there is a set of non-overlapping solutions $\mathrm{ACQ}_i(\boldsymbol{p}^*)$ that demand all the goods with nonzero prices:

$$\forall i, i'. \mathrm{ACQ}_i(\boldsymbol{p}^*) \cap \mathrm{ACQ}_{i'}(\boldsymbol{p}^*) = \emptyset,$$
$$\bigcup_i \mathrm{ACQ}_i(\boldsymbol{p}^*) = \{x_i \in \mathcal{X} \mid p_i > 0\}.$$

Price vectors satisfying this definition exist under well-defined conditions [Mas-Colell et al., 1995], however, these are generally violated for discrete goods and especially in the presence of complementarities. Nevertheless, given a model of the agent population and their valuations, one can search for prices that meet or approximate the equilibration using iterative methods, and this provides a practical approach to generating initial price predictions. As noted above, a strategy that employed this technique in the TAC Travel game [Cheng et al., 2005] was quite effective, perhaps surprisingly so given the violation of preconditions for equilibrium existence and the fact that this method refrains from exploiting available historical price information.

A second broad approach to price predictions is to simulate or otherwise model the way prices are formed through bidding and auctions. For example, if our aim is to predict the prices for a CDA scenario, we could collect observations of simulated runs of the mechanism, sampling from the distribution of agent valuations. The sample average prices, or the sample distribution of prices, can then serve as the prediction, or combine with other information in a prediction estimate. This approach also requires that we assign strategies to agents, for instance in the CDA context we might adopt HBL or other strategies shown in analysis to be effective in equilibrium (see Section 4.4.2). For one-shot simultaneous markets, we could select from the strategies introduced in Section 5.2. Recall that these strategies in turn take price predictions as input, so we have a problem of bootstrapping. The initial setting can use any reasonable baseline prediction, for example one developed using competitive analysis or other model-based approach.

The circularity noted above suggests a third, iterative approach to prediction. Starting from some initial prediction, we compute (e.g., through simulation) the prices that would result from agents implementing plausible strategies taking that prediction as input. These price outcomes then form the basis for a new prediction, which we can similarly employ as input to bidding strategy, thus generating a new outcome. A fixed point of this process is what we call *self-confirming prices*, that is, prices that when used as inputs by prediction-based bidding strategies, turn out to be correct. More formally, if PP is a strategy that employs price prediction, we denote by PP($\boldsymbol{p}$) the strategy that plays PP with input prediction $\boldsymbol{p}$. Let $\mathcal{E}(\mathsf{S})$ denote the distribution of market prices resulting when all agents play strategy S in market environment $\mathcal{E}$.

Definition 5.9   Self-Confirming Prices   A price vector $\boldsymbol{p}^*$ is *self-confirming* with respect to PP in market environment $\mathcal{E}$ if it represents the expected price outcome given strategies using that prediction:

$$p_i^* = \mathbb{E}_{\mathcal{E}(\mathsf{PP}(\boldsymbol{p}^*))}[p_i], \ i \in \{1, \ldots, m\}.$$

A price distribution F^* is *self-confirming* with respect to PP in market environment $\mathcal{E}$ if

$$F^* = \mathcal{E}(\mathsf{PP}(F^*)).$$

Observe that a competitive equilibrium can be viewed as a vector of self-confirming prices with respect to the TargetPrice strategy in a first-price auction mechanism. In a competitive equilibrium, the target sets are disjoint, so when each agent bids the predicted price for its respective target bundle, that is indeed the outcome. Just as for competitive equilibria, self-confirming prices or distributions may not exist in many environments. We may nonetheless appeal to the principle of self-confirmation by generating price predictions through the iterative process sketched above. That is, starting from a baseline prediction and a strategy assignment, refine the prediction on each step by simulating the strategies in a sample from the specified market environment. If the process converges (within some tolerance), or a maximum time or other termination criterion is reached, we return the closest to self-confirming prices found to that point as the candidate prediction.

This iterative method for deriving self-confirming prices can be applied for bidding in any auction mechanism, as long as we have a strategy based on price prediction and a model of the market environment. We evaluated this approach in a study of bidding in simultaneous SPSB auctions [Yoon and Wellman, 2011], using bundle valuations based on scheduling problems as in Example 5.3. In this environment, the iterative method reliably produced price distributions that are approximately self-confirming, with respect to the full range of bidding strategies discussed in Section 5.2 above. In an empirical game analysis, strategies from the BidEvaluator family were clearly best; in fact all others were dominated. Although not competitive as standalone strategies, the heuristic methods are still instrumental as components of BidEvaluator. Particular effective were BidEvaluator(StraightMV) and BidEvaluator(TargetMV*), which use StraightMV and TargetMV*, respectively, to generate bid candidates and self-confirming price predictions for evaluation. Also effective were hybrid BidEvaluator approaches, which used different or even multiple seed strategies for these purposes.

5.4   SIMULTANEOUS ASCENDING AUCTIONS

In iterative market mechanisms, information received as the market proceeds can inform an agent's subsequent bidding decisions. For instance, we have seen in preceding chapters how strategies for open-outcry auctions or CDAs can exploit information revealed dynamically over an auction process. Such information provides evidence about underlying valuations of other bidders, and can be used to sharpen predictions about the market outcomes.

A *simultaneous ascending auction* (SAA) [Cramton, 2005] is an iterative market mechanism which allocates a set of related goods via separate, parallel English auctions for each good. This is characteristic of many real-world scenarios, such as suppliers participating in concurrent procurement auctions run by different purchasers. Simultaneous ascending auctions have also been employed in prominent spectrum auctions in the U.S. and other countries, and variants applied to power markets and other explicitly designed trading environments [Milgrom, 2003]. Many of the strategic issues presented by SAAs apply as well whenever there are concurrent markets for interrelated goods.

5.4.1   SAA OPERATION

The SAA mechanism comprises m separate auctions, one for each good, that operate over multiple rounds of bidding. The bidding is synchronized so that in each round each agent submits a bid in every auction in which it chooses to bid. At any given time, the BID price on good i is BID_i, defined to be the highest bid b_i received thus far, or zero if there have been no bids. The BID quote along with the current winner in every auction is announced at the beginning of each new round. To be admissible, a new bid must satisfy $b_i \geq ASK_i = BID_i + \delta$. We take the bid increment $\delta = 1$ for this SAA discussion. [4] If an auction receives multiple admissible bids in a given round, it admits the highest, breaking ties randomly. An auction is *quiescent* when a round passes with no new admissible bids. When every auction is simultaneously quiescent they all close, allocating their respective goods per the last admitted bids.

An agent's *current information state*, $\boldsymbol{B}$, comprises the current BID prices, along with a bit vector indicating which of the goods the agent is currently winning. In general, an agent's bidding strategy maps the *history* of information states to bids. For the present discussion, we consider a more limited space of strategies that condition behavior on only the current information state. A strategy is then a mapping from this information state to its bid for each of the m auctions. Submitting an inadmissible bid (e.g., $b_i = 0$) is equivalent to not bidding. An agent's payoff (surplus), $\sigma(X, \boldsymbol{p})$, is governed by the auction outcomes, namely, the set of goods it wins, X, and the final prices, $\boldsymbol{p}$ (5.2).

5.4.2   SAA STRATEGY

How should an agent bid in a simultaneous ascending auction?

5.4.2.1   Straightforward Bidding

As a starting point, consider the so-called *straightforward bidding* (SB) strategy, [5] which bids for the set of goods that would provide the greatest surplus at the current prices. By current prices, the agent considers that it could obtain goods it is currently winning at the BID price, and other goods at the ASK price. We formalize this notion using a *perceived-price* function, ρ:

$$\rho_i^{SB}(\boldsymbol{B}) = \begin{cases} BID_i & \text{if winning good } i, \\ BID_i + 1 & \text{otherwise.} \end{cases}$$

SB determines a target bundle by optimizing surplus at perceived prices $\rho \equiv \rho^{SB}$:

$$X^* = \arg \max_X \sigma(X, \rho(\boldsymbol{B})), \tag{5.4}$$

and submits bids at the ASK price for those goods in X^* which it is not currently winning.

To illustrate, suppose all three agents play SB in an SAA for the scheduling scenario of Example 5.3. The specific result depends on tie-breaking sequence, but one representative possibility is that agent 1 ends up with $\{1\}$ at price 7, agent 3 with $\{2\}$ at price 5, and agent 2 with $\{3\}$ at price 4. The latter two agents realize positive gains, but agent 1 suffers a loss. Under SB, agent 1 bids for its entire bundle until the prices exceed valuation, and ultimately gets stuck with a good it cannot use.

---

[4]This is without loss of generality, allowing for scaling of agent values.
[5]We adopt the terminology introduced by Milgrom [2000]. The same concept is also referred to as "myopic best response", "myopically optimal", or "myoptimal" [Kephart et al., 1998].

Once again, we observe a disparity between agents with diminishing and increasing marginal value. For the former, SB completely resolves the problem of exposure in SAAs, [6] as ascending prices ensure that a price change in one good can only increase the value of other goods the agent is winning. This is not the case with complementary preferences, and indeed the SB strategy is oblivious to the exposure problem faced by agents with such valuations.

5.4.2.2 SAA Bidding with Price Prediction

In this example, if agent 1 could have anticipated that it would be unable to profitably obtain the bundle, it might have refrained from bidding altogether. One way to express such anticipation is through price predictions, of the same sort discussed in sections above. For example, with a predicted price vector of $\langle 7, 5, 4 \rangle$, agent 1 playing one of the targeted bidding strategies in the one-shot setting would not bid positively for any good.

We can similarly exploit such predictions in dynamic markets like SAA, through a straight-forward extension of straightforward bidding. Let p be a vector of predicted prices. We can define the perceived prices at information state B with respect to prediction p as follows:

$$\rho_i^p(B) = \begin{cases} \max(p_i, BID_i) & \text{if winning good } i, \\ \max(p_i, BID_i + 1) & \text{otherwise.} \end{cases}$$

The price prediction strategy PP(p) then behaves just like SB, with ρ^p substituted for ρ in (5.4). [7] Note that straightforward bidding is the special case of price prediction with the predictions all equal to zero: SB = PP(0). If the agent underestimates the final prices, it will behave identically to SB after the prices exceed the prediction. If the agent overestimates the final prices, it may stop bidding prematurely.

Employing predictions in the form of probability distributions is conceptually similar, with optimization based on expected surplus rather than simple surplus. Let F denote the predicted distribution of final auction prices. As with the point predictor, we restrict the updating in our distribution predictor to conditioning F on the fact that prices are bounded below by the BID prices. Let $\Pr(p \mid B)$ be the probability, according to F, that the final price vector will be p, conditioned on the information revealed by the auction, B. Then, with $\Pr(p \mid \varnothing)$ as the pre-auction initial prediction, define:

$$\Pr(p \mid B) \equiv \begin{cases} \dfrac{\Pr(p \mid \varnothing)}{\displaystyle\sum_{q \geq BID} \Pr(q \mid \varnothing)} & \text{if } p \geq BID \\ 0 & \text{otherwise.} \end{cases} \tag{5.5}$$

---

[6]The observation holds more precisely for value functions exhibiting *gross substitutes*, where an increase in price of one good cannot cause another to be dropped from an optimal acquisition [Lehmann et al., 2006, Milgrom, 2000]. Gross substitutes is entailed by diminishing marginal value at *all* prices. During an SAA, an agent is immune from exposure as long as diminishing MV holds for prices greater than or equal to current prices.

[7]One may obtain a broad family of bidding strategies through variations on the perceived-price function ρ [Wellman et al., 2008], including approaches not based explicitly on price prediction.

(By $\boldsymbol{x} \geq \boldsymbol{y}$ we mean $x_i \geq y_i$ for all i.) For (5.5) to be well defined for all reachable information states, we must ensure that all possible price vectors have positive initial probability.

For example, suppose an agent's initial prediction in our running example is that prices are independent and uniformly distributed: $U[0, 12]$, $U[0, 10]$, and $U[0, 4]$ for the respective goods. At a current BID quote of $\langle 5, 2, 1 \rangle$, the agent's posterior distributions would be $U[5, 12]$, $U[2, 10]$, and $U[1, 4]$.

An agent can further use distribution information to take account of the fact that goods it wins based on existing bids represent *sunk costs*. If an agent is currently not winning a good and bids on it, then the expected incremental cost of winning the good is the expected final price, calculated with respect to the distribution F. If the agent is currently winning a good, however, then the expected incremental cost depends on the likelihood that the current bid price will be increased by another agent, so that the first agent has to bid again to obtain the good. If, to the contrary, it keeps the good at the current bid, the full price is sunk (already committed) and thus should not affect incremental bidding. Based on this logic, we define $\Delta_i(\boldsymbol{B})$, the expected *incremental* price for good i.

First, for simplicity, let us consider only the information contained in the vector of marginal distributions, $(F_1, \ldots, F_m)$, as if the final prices were independent across goods. Define the expected final price conditional on the most recent bid prices:

$$\mathbb{E}_F(p_i \mid \boldsymbol{B}) = \sum_{q_i \geq 0} \Pr(q_i \mid BID_i)q_i = \sum_{q_i \geq BID_i} \Pr(q_i \mid BID_i)q_i.$$

The expected incremental price depends on whether the agent is currently winning good i. If not, then the lowest final price at which it could win is $BID_i + 1$, and the expected incremental price is simply the expected price conditional on $p_i \geq BID_i + 1$,

$$\Delta_i^{\mathrm{L}}(\boldsymbol{B}) \equiv \mathbb{E}_F(p_i \mid p_i \geq BID_i + 1) = \sum_{q_i \geq BID_i + 1} \Pr(q_i \mid p_i \geq BID_i + 1)q_i. \tag{5.6}$$

If the agent is winning good i, then the incremental price is zero if no one outbids the agent. With probability $1 - \Pr(BID_i \mid BID_i)$ the final price is higher than the current price, and the agent is outbid with a new bid price $BID_i + 1$. Then, to obtain the good to complete a bundle, the agent will need to bid at least $BID_i + 2$, and the expected incremental price is

$$\Delta_i^{\mathrm{W}}(\boldsymbol{B}) = (1 - \Pr(BID_i \mid BID_i)) \sum_{q_i \geq BID_i + 2} \Pr(q_i \mid BID_i + 2)q_i.$$

The vector of expected incremental prices is then defined by

$$\Delta_i(\boldsymbol{B}) = \begin{cases} \Delta_i^{\mathrm{W}}(\boldsymbol{B}) & \text{if winning good } i, \\ \Delta_i^{\mathrm{L}}(\boldsymbol{B}) & \text{otherwise.} \end{cases}$$

The distribution-predicting agent then plays the perceived-price bidding strategy (5.4) with $\rho(\boldsymbol{B}) \equiv \boldsymbol{\Delta}(\boldsymbol{B})$. We denote the strategy of bidding based on a particular distribution prediction by $\mathsf{PP}(F^x)$, where x labels various pre-auction distribution predictions, $F(\varnothing)$.

For instance, return to our running example, where agent 1 started with prediction $\langle U[0, 12], U[0, 10], U[0, 4]\rangle$, and has reached a state where it is winning goods 1 and 3 at current prices $\langle 5, 2, 1\rangle$. The agent's vector of expected incremental prices is

$$\Delta(\boldsymbol{B}) = \langle \Delta_1^W(\boldsymbol{B}), \Delta_2^L(\boldsymbol{B}), \Delta_3^W(\boldsymbol{B})\rangle$$
$$= \langle (7/8)9.5, 6.5, (3/4)3.5\rangle = \langle 8.31, 6.5, 2.63\rangle.$$

Based on these anticipated incremental costs to obtain its bundle, agent 1 would refrain from bidding further on these goods (and hope to be outbid on those it was currently winning).

5.4.2.3 Demand Reduction

Price-prediction strategies directly address the issue of *exposure*, identified in Section 5.1 as the primary implication of interdependent markets for bidding strategy. For agents with complementary preferences, the exposure risk is that of failing to obtain the full bundle justifying the bids placed for goods obtained. The difference in value can be drastic, thus exposure constitutes one of the most important strategic issues for bidding in SAAs with complements. When goods are substitutes, in contrast, ascending prices only increase demand for other goods, and thus the SAA mechanism provides a shield against exposure. In such circumstances, the second implication of interdependence—*influence* across markets—comes to the fore.

The issue of influence is that an agent's bid in one market may affect the prices it faces in other markets. To illustrate, consider the extreme case of *perfect substitutes*, where goods are essentially identical. By bidding to obtain one good, the agent exerts positive force on the prices of the others. This is precisely the issue encountered for bidding in multiunit auctions, discussed in Section 4.2.2. As seen there (in a call market context), an agent demanding multiple units may be able to obtain fewer units at a lower per-unit price, and this can be beneficial even if the price is below the marginal value of a unit.

For example, consider an agent who views the goods as perfect substitutes, valued linearly at 100 each. Suppose that if it bids to obtain eight, the price will be 88, but if it goes for nine the price will be 90 (assuming other agents also view them as substitutes, the increased demand will raise prices for all goods). The surplus for buying eight goods ($8 \times (100 - 88) = 96$) in this case is greater than that for buying nine ($9 \times (100 - 90) = 90$).

In a demand reduction approach, the agent accounts for the influence of its own bids on prices by choosing to bid for fewer goods or at lower prices than otherwise called for by its baseline strategy. There are many possible ways to implement demand reduction in a bidding strategy; here I outline a simple and direct strategy suitable for bidding in SAAs.

The demand reduction strategy, DR, applies specifically to the case of perfect substitutes. [8] Let $\kappa \geq 0$ be a parameter defining the degree of demand reduction. An agent playing strategy $DR(\kappa)$ bids the ASK price on the lth cheapest good as long as it is not winning that good, and its marginal surplus is at least $\kappa(l - 1)$. In other words, the agent considers the goods in order of price,

---

[8]The basic idea can be generalized to imperfect substitutes, with care to distinguish the goods.

adding the lth good to its bundle until the marginal value μ_l drops below $ASK + \kappa(l-1)$. The DR strategy family is a simple way of capturing the intuitions of the demand-reduction literature: bidders should shade their bids, and the amount of shading increases with the number of winning goods [Ausubel and Cramton, 2002].

Formally, define DR's perceived price of the good with the lth lowest myopically perceived price:

$$\rho_l(\boldsymbol{B}) \equiv \begin{cases} BID_l + \kappa(l-1) & \text{if winning the good,} \\ BID_l + 1 + \kappa(l-1) & \text{otherwise.} \end{cases} \tag{5.7}$$

Agent DR(κ) plays the perceived-price bidding strategy using this $\rho(\boldsymbol{B})$. Note that $\rho(\boldsymbol{B})$ as defined by (5.7) assumes that the goods are indistinguishable. We use the subscript l instead of i to emphasize that each good is labeled by its myopic price rank order rather than by the auction selling it.

5.4.3   EMPIRICAL GAME-THEORETIC ANALYSIS

To evaluate candidate strategies for SAA bidding, we conducted an extensive EGTA study over a range of valuation environments [Wellman et al., 2008]. Since the implications of interdependence in the form of substitutes are qualitatively different from those of complements, we treat these cases separately.

For environments with complementary preferences, we constructed environments where agent valuations derive from scheduling tasks, as in Example 5.3. In each of these we evaluated a suite of strategies, representing the plausible candidates previously proposed in the literature. These include SB along with explicit price-prediction strategies, with predictions in the form of points and distributions, generated through diverse approaches such as those discussed in Section 5.3. Environments varied in number of agents ($3 \leq N \leq 8$) and number of goods ($3 \leq m \leq 7$), and in the distributions over number of slots (goods) that agents need to perform their task. Although the specific results varied across environments, overall we found strong support for PP(F^{SC}), the strategy based on self-confirming distribution predictions. In almost all of the environments tested, the strategy profile with all agents playing PP(F^{SC}) was a Nash equilibrium, or approximately so. Although some room for improvement exists, available evidence indicates that self-confirming price prediction is an effective approach to dealing with the exposure problem. For SAAs with complementary preferences, PP(F^{SC}) can thus be viewed as the "reigning champion" strategy, that is, a provisional best choice pending identification of alternatives superior in particular environments.

For the case of substitute preferences, the recommendation is not so categorical. In EGTA studies with homogeneous goods (i.e., perfect substitutes), strategies from the DR family emerge best, in competition with SB and price-prediction strategies. Efforts to devise sophisticated predictions of own-price effects (by analogy to self-confirming price prediction) have thus far not proven superior to simple demand reduction approaches. Implementing DR(κ) requires choosing a value for the demand reduction parameter κ, whose optimal setting may vary across environments.

5.5   STRATEGY LESSONS

Let us summarize once again the key strategic insights about simultaneous interdependent markets (one-shot or ascending) presented in this chapter.

1. Interdependent markets present bidders with an *exposure problem*: they must make decisions in one market before they know the outcomes of other relevant markets. Effective strategies account for uncertainty in balancing exposure risks with the benefits of acquiring desired bundles.

2. *Marginal value* provides a measure of value for an individual good, given assumptions about the other goods obtained or available through buying opportunities.

3. *Complementary goods* pose a particularly difficult exposure problem. Complementarity in preference is characterized by marginal value that is increasing in the set of other goods held. When marginal value is *not* increasing, the marginal values provide an effective guide to which goods are part of an *optimal acquisition*.

4. *Price prediction* provides a basis for making exposure tradeoffs, and this is a central component of strategies for bidding in interdependent markets. There are a wide variety of techniques for price prediction, based on machine learning and analytical models.

5. There is a rich space of strategies that employ price predictions and concepts like marginal value to generate bids for simultaneous interdependent markets. Particularly flexible and effective is the BidEvaluator approach, which uses heuristics to seed candidates for evaluation with respect to predicted outcomes.

6. *Self-confirming price predictions* are distributions over prices that turn out to be accurate when all agents use them in a price-predicting bidding strategy. They can be derived through simulation models, and can be applied even when no historical price information is available.

7. *Simultaneous ascending auctions* are an iterative form of interdependent markets. The *perceived-price strategy* is a general approach to SAAs that encompasses straightforward (myopic) bidding as well as methods that incorporate price predictions.

8. For SAAs with complementary preferences, the exposure problem is paramount, and bidding based on self-confirming price predictions is an effective strategy.

9. For *substitute preferences* (absence of complements), the dominant strategic issue in SAAs is the influence of behavior in one market on outcomes in others. Simple *demand-reduction strategies* provide an effective way to modulate influence in such circumstances.

As for continuous double auctions, canonical game-theoretic models of interdependent markets resist analytic solution. As such, we cannot present the last word on bidding in simultaneous auctions, for either iterative versions such as SAA or even the one-shot case. Nevertheless, the taxonomy of strategies presented here provides a starting point for designing trading agents for a wide range of scenarios featuring interdependent markets.

5.6   BIBLIOGRAPHIC NOTES

Despite the practical importance of simultaneous interdependent markets, there is not a great deal of auction theory literature on the subject. One early exception was the analysis by Engelbrecht-Wiggans and Weber [1979] of an example simultaneous SPSB auctions for substitute goods. Gerding et al. [2008] also analyze a case of perfect substitutes, and algorithmically derive approximate equilibria where bidders bid high (probabilistically close to true value) in one auction and low in the rest. Krishna and Rosenthal [1996] consider simultaneous SPSB auctions when some bidders exhibit complementary preferences and the rest value only a single unit.

The pivotal distinction between complementary and substitute preferences over good bundles is widely recognized. A detailed taxonomy of qualitative preference classes is presented by Lehmann et al. [2006]. The exposure problem induced by complementarity in simultaneous auctions came to prominence in discussion of the FCC spectrum auctions in the mid-1990s [Bykowsky et al., 2000, Cramton, 1997].

Dealing with multiple markets has been a theme in AI work on trading agents [Byde et al., 2002, Greenwald and Boyan, 2001, Wellman and Wurman, 1999]. Greenwald and Boyan [2004] framed the problem of bidding across interdependent markets given probabilistic price predictions. Follow-on work [Greenwald et al., 2009, Wellman et al., 2007] formalized this bidding problem in decision-theoretic terms, and established properties of optimal bidding strategies given the assumption that bids do not affect other-agent behaviors. Further experimental comparison was performed by Greenwald et al. [2010]. These works introduced the taxonomy of heuristic bidding strategies presented in Section 5.2.

Self-confirming price predictions were originally defined in the context of simultaneous ascending auctions by MacKie-Mason et al. [2004]. The FCC auctions drew considerable attention to strategic issues for SAA mechanisms [Milgrom, 2000]. Much of the research on SAA bidding strategy focuses on issues of demand reduction perceived as pivotal in that domain [Reitsma et al., 2002, Weber, 1997]. Our discussion of SAA bidding follows the treatment of Wellman et al. [2008].

This chapter does not include a detailed treatment of strategy for interdependent auctions operated sequentially. As suggested by our discussion of related problems—shopping on eBay and trading multiple units in CDAs—the natural approach to reasoning about such situations is from the last stage backward. Sequential bidding problems can be more precisely formulated in terms of dynamic programming [Boutilier et al., 1999] or Markov decision processes [Greenwald and Boyan, 2004]. Some recent works have considered the use of options [Juda and Parkes, 2009, Mous et al., 2010] as a way of managing exposure in sequential auctions.

CHAPTER 6

Conclusion

In this volume we considered a variety of issues bearing on trading agent strategy. Their relative salience depends on many contextual factors, which we can group into three roughly orthogonal dimensions:

1. Individual auction rules, which can be organized [Wurman et al., 2001] according to how they govern:

   - *bidding*: what bids are eligible and when;
   - *pricing*: how trade prices are determined as a function of winning bids;
   - *timing*: when bids are matched; and
   - *information*: what (if any) state information is revealed by the auction in the interim during the course of bidding.

2. Dependencies across agents:

   - *valuations*: how each agent's preferences depend on others'; and
   - *signals*: what information each agent has about the private values of the rest.

3. Interactions across markets for different goods (or multiple units of the same good):

   - *preferences*: degree of complementarity or substitutability among goods, and
   - *temporal order*: whether markets operate sequentially, simultaneously, or some combination.

Design of a trading agent starts with understanding the market environment in which it will operate, which includes locating it on the dimensions above. Accounting for multiple factors entails putting together concepts described separately in the preceding chapters. We thus conclude the text with an attempt at integrating the ideas into a comprehensive approach to trading agent design.

6.1 TRADING AGENT ARCHITECTURE

It is common in agent design to describe, typically in block-diagram form, the high-level organizational structure of the agent's decision-making process. The term *agent architecture* refers to such a high-level design, specifying the fixed processes and information flows employed by agents that instantiate this architecture. These agents may also invoke other processes for particular application domains; the architecture defines the invariant parts of the overall design.

One invariant prescription for trading agents is that they include an arbitrage layer. The prerequisite for automating arbitrage (see Section 1.3) is a precise definition of identity relationships among goods across markets, such that we can calculate the translation of one combination of goods to another. Given such a definition, we can recognize a profit opportunity based on prevailing prices. The recipe for arbitrage identification, therefore, is to connect opportunity recognition with frequent monitoring of prices at relevant markets. Since trading strategies geared to fundamental objectives will invariably entail such monitoring anyway, including the arbitrage layer may be accomplished as a matter of course.

In trading agents designed for realistic environments, many of the details of strategy and technique lie in the non-invariant areas that differ across application domains. Nevertheless, we can identify some common processes and flows, which can be described in terms of the basic bidding cycle at the core of a trading agent we would design for almost any complex domain.

6.2 BASIC BIDDING CYCLE

For any trading environment more complex than a single one-shot auction, a strategy must interact with the market over time, taking actions in the form of bids, receiving information and possibly results, submitting new or revised bids, and so on. The agent thus can be viewed as operating over a cycle of action and observation, until its objectives are satisfied or the market mechanism terminates. The cycle may be periodic or continual, or even irregular, depending on the schedule by which the market reveals information.

Based on the common structure found in many of the strategies we have discussed for complex markets, we can elaborate somewhat on the computational activity of the agent during each pass through the cycle. The basic trading agent bidding cycle is depicted in Figure 6.1. We discuss each of the steps in turn, illustrating how they are manifest in a complex trading agent, DeepMaize [Kiekintveld et al., 2006], designed and implemented at the University of Michigan for the TAC Supply Chain Management (TAC/SCM) scenario [Collins et al., 2010]. Although a detailed treatment of the SCM environment is beyond the scope of this volume, presenting the basic cycle in the context of a rich trading scenario demonstrates how domain-specific computations map to the generic stages defined here.

6.2.1 UPDATE MARKET STATE

The first step prepares for bidding by gathering the latest information from the environment and incorporating it into the agent's model of market state. Observations include reports on transactions (by this agent or others, individually or in aggregate), and price quotes provided by auctions to reveal summaries of their order books. In simple market configurations, mere recording suffices to incorporate the information. We have also seen somewhat more elaborate incorporation operations,

Figure 6.1: Basic bidding cycle.

such as the update of various market statistics for CDAs (Section 4.3), or the updates of information state based on price quotes in SAAs (Section 5.4).

For richer scenarios, the update may apply over a broader model of environment state. In the TAC/SCM scenario, agents represent manufacturers, and simultaneously trade in markets for components and finished goods, each market taking bids and generating transactions daily over a series of 220 days. Each agent receives daily feedback about its own bidding—offers in response to its requests for quotes (RFQs) from suppliers, and information about which bids for customer RFQs it won—and periodic market reports with aggregated information about other-agent transactions. In addition, the pattern of supplier deliveries and the volume of new customer demand provides key information about the market environment. DeepMaize attempts to synthesize all of this information to update its models. It employs models of supplier request response to estimate current uncommitted capacity levels, and a stochastic model of customer demand trends to perform a Bayesian update of the key demand parameters.

6.2.2   PREDICT PRICES

The second step of the bidding cycle is price prediction. The construction and use of explicit price predictions was introduced explicitly above for dealing with interdependent markets (Section 5.3). Predictions may take the form of point prices, or price distributions, independent or jointly across markets, and may or may not account for the impact of the agent's own demand (i.e., own-price effects). Even for single markets, predicting prices is an essential technique. Recommended strategies for first-price auctions (Section 3.2.1) and auctions with interdependent values (Section 3.3) considered the probability of winning at various bids, which is tantamount to probabilistic price prediction. CDA strategies from the potent HBL family (Section 4.3.4) rely on explicit price prediction as well.

The central role of price prediction in agents developed for TAC scenarios is also discussed in Section 5.3. In TAC/SCM, predicting agents must deal with two categories of markets—for components and finished goods—that differ qualitatively in market mechanism, information available, and supply and demand characteristics. Moreover, since agents may choose when to purchase and sell goods in a dynamic market, they are interested in predicting prices at all future points in time. To account for these various facets, the TAC/SCM Prediction Challenge [Pardoe and Stone, 2010a] evaluated agents on their accuracy in four distinct prediction tasks: for components and finished goods, both in the current period and at a point 20 days hence.

To predict prices of finished goods, DeepMaize considers features based on its market state estimation (previous step), such as underlying customer demand parameters, as well as some raw statistics on recent observed prices and trends. It matches these features to a large database of previous SCM scenario snapshots, retrieving the most similar cases, or *nearest neighbors*. Its probabilistic prediction for the current price as well as any point in the future is then constructed from a weighted distribution of prices across these neighbors, adjusted online during a game scenario based on observed performance [Kiekintveld et al., 2009]. Component price predictions are actually more complex, as there is a separate price for every combination of supplier, order date, and delivery date. DeepMaize employs a variety of methods based on interpolation and regression, with the specific method chosen for tournament play based on empirical game analysis.

Where state estimation ends and price prediction begins is something of an arbitrary distinction, and does not really matter in any fundamental way. Nevertheless, from the perspective of agent architecture it is often quite helpful to decompose these steps into separate modules, for example to facilitate experimentation with alternative price prediction algorithms that employ the same basic model of market state.

6.2.3   OPTIMIZE TARGET BUNDLE

Given a prediction of market prices, the agent can decide what bundle of goods it would like to buy and sell in the market. The third step of the basic bidding cycle is to make this decision, at least provisionally pending resolution of market uncertainty. For agents that only buy goods, we term this the acquisition problem (Definition 5.6), and generalized versions that consider selling as well are analogous. The significance of identifying a target bundle may vary across bidding strategies. As discussed in Section 5.2.1, targeted bidding strategies submit bids only for goods in the target set, whereas others (e.g., StraightMV) may submit bids outside this set to hedge against cases where prices vary far from the expected.

More generally, a trading agent constructs bids according to some overall plan for buying and selling over time. In a manufacturing scenario like TAC/SCM, the agent should exploit its freedom about the precise timing and identity of procured components, but having a rough tentative schedule for acquiring components is a useful guide in formulating bids. DeepMaize constructs plans in detail for short-term procurement and sales, and at a coarser level for farther time horizons (see Figure 6.2). The specific plan for the current day details which components it will attempt to procure from which supplier, and which customer RFQs it will bid on. The tentative longer-term plans include rougher aggregate estimates. For components, it specifies the quantities of each type of component to acquire by each time, without committing to particular suppliers or delivery dates. The sales plan is likewise an aggregated projection, specifying overall quantities of finished goods it expects to ship, but not broken down into particular orders by date. These projections are based on optimizations of a relaxed version of the problem, which considers for example the overall factory capacity, but does not work out a specific factory schedule to verify that it is feasible. As we plan further out the information available becomes more uncertain and projections less reliable, and thus precision is less important.

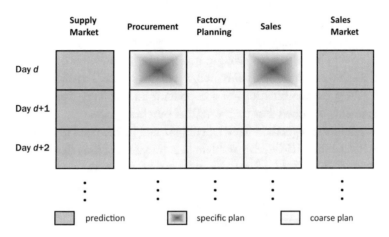

Figure 6.2: On day d, DeepMaize predicts prices for supplies and sales for all dates, constructs a specific plan for procurement and sales this day, and forms coarse-grain plans for factory scheduling and future sales and procurement.

The purpose of building a coarse-grained plan for the future is to provide a principled basis for evaluating candidate near-term plans. The value of obtaining a component now depends on our uses for it now or in the future, and our alternative opportunities for obtaining it later. Similarly, the expected profit from selling a finished good depends on our expected costs for replacing the component inventory, or opportunity cost of forgoing a future sale. DeepMaize performs these calculations explicitly, assessing the marginal value of having incremental units of each component and each finished good, at all times in the future. These values provide a convenient way to manage the interaction among major modules of a complex agent [Kiekintveld et al., 2006]. For example, the DeepMaize sales module can optimize profits taking as given information about the marginal costs of components. Similarly, the factory scheduler can formulate its own optimization problem in terms of given values for producing finished goods at any point in time.

6.2.4   COMPUTE BIDS

The final step in our basic bidding cycle is to compute bids and submit them to the respective market mechanisms. We discussed a variety of heuristic strategies for constructing bids, given target bundles and an ability to compute marginal values with respect to price predictions (Section 5.2). For one-shot markets (or treating one iteration of the cycle in isolation), we can formulate an optimal bidding problem in these terms, and employ the BidEvaluator strategy or other technique to derive an approximate solution. The optimization here is relative to the assumptions of our price-prediction method, and the other steps of the cycle.

For DeepMaize, the marginal values computed in the process of constructing short- and long-term plans (previous step) can provide a direct guide for bidding. For procurement, bidding comprises both generation of RFQs, and decisions for offers received in response to previous RFQs. For both functions, DeepMaize uses the projected production schedule to identify required components, and associates these with marginal values as a function of when they arrive in inventory. It searches over the combinations of supplier offers (received in response to the previous day's RFQs) to accept the set maximizing the value of components obtained, net of offer price. To solicit offers for the next day, it generates new RFQs for combinations of components and delivery dates that would maximize the value of components obtained, less their *predicted* purchase price.

Bidding to sell finished goods is likewise complicated by the need to consider many orders at once. Given a set of customer RFQs, the problem is to find a set of bids that maximizes expectation revenue minus cost of goods sold, where the costs are based on values from the projected plan (including expected orders won this day), and expectation is based on the probability of winning combinations of orders at candidate prices. DeepMaize uses a form of BidEvaluator that performs a local search in the space of reasonable bids, bounded below by the marginal cost of a finished good.

6.3   TRADING AGENT RESEARCH

The Introduction to this volume enumerated many forces behind the impetus to automate trading activity, such as proliferation of electronic interfaces, and the need to monitor many information channels quickly and cheaply, and respond rapidly to opportunities. These forces are not likely to abate anytime soon, and so it seems an easy call to predict that the population of trading agents will continue to multiply. The precise scope and extent of this automation is much harder to forecast, so I will just note that I envision no fundamental limits. The most difficult barriers to trading agent deployment are that of expressing in an efficient and reliable way the designer's preferences (e.g., costs and valuations of trading outcomes) in agent-usable form. Understanding of explicit and implicit objectives, rather than any aspect of trading agent strategy itself, is where humans are most apt to remain superior to autonomous agents in the near future at least.

The principles for trading agent design expounded here are the product of decades of analysis and experience from economic and AI research. As for any field of active inquiry, however, accepted conclusions are subject to revision and refinement as our knowledge accumulates. Whereas the theorems of auction theory are presumed eternal, their implications for practical trading agent design are anything but. Continual evaluation of strategy ideas through theory and simulation, in generic and domain-specific settings, through models and field trials, will undoubtedly shed more light on trading agent problems in the years to come. Research naturally generates questions as well as answers, so even our conception of what the key issues are is quite likely to evolve over time.

Although the specific paths that the research enterprise will take are unpredictable, we can reasonably anticipate progress in several important areas.

Analysis of more elaborate market mechanisms.   As recounted herein, auction theory has identified equilibrium strategies for basic mechanisms, under a range of valuation models. The theory also addresses many variations, including some forms of multiunit and some simple cases of sequential auctions. Work continues to extend these results, pushing the complexity envelope of market environments understood in these terms.

New champion strategies for complex market environments.   Where general theoretical analysis remains intractable (e.g., CDAs or simultaneous markets), experimental comparisons still provide evidence about the best strategies developed to date. Such results are provisional, in that they are always subject to overthrow by new champions put forth by the research community. Although one might prefer once-and-for-all results, tracking the survivors of such a competitive research process provides a next-best approach to vetting strategy ideas for important classes of market mechanisms. Over time, the process yields a family tree of heuristic trading strategies, amenable to systematic exploration and classification.

Strategies for advanced auction mechanisms.   The automation of markets also opens opportunities for deploying innovative mechanisms that directly address complex market challenges. The prospect of combinatorial auctions—mechanisms for direct allocation of bundles—has drawn particular attention from the academic research community [Cramton et al., 2005]. Also offering intriguing capabilities are *multiattribute auctions*, which determine features as well as parties of trades through a bidding process [Bichler, 2001, Parkes and Kalagnanam, 2005]. Although these classes of mechanism are not yet commonly employed, when and if they become more prevalent they will certainly become ripe targets for trading agent design.

Learning and price prediction.   We are yet in the early days of understanding what market features and machine learning techniques are most effective for constructing forecasts of market prices. To what extent there are useful domain-independent principles for price prediction is at present an open question in trading agent research.

Domain-specific trading strategies.   I have strived in this text to focus on general principles of trading agent design, applicable regardless of the particular market domain. Specialized trading realms, however, do offer opportunities to gather and apply domain-specific knowledge. Financial trading, of course, is the most significant sector with a highly developed body of technique from research and practice. Established methods from this domain must be mastered in order to produce viable agents for financial trading. Analogous specialized knowledge is likely to develop in emerging trading domains, for example energy markets. The importance of special technique is proportional to the domain-specific nature of the mechanisms and other environmental elements in use in that domain. For example, the idiosyncratic structure of auctions for Internet search keyword advertising [Lahaie et al., 2007] combined with the magnitude of this market renders the sponsored-search domain a prime candidate for research in specialized trading agent methods.

Trading agent design methodology.   Let us conclude with a meta-level point. Ultimately, the field of trading agent design advances not just by identifying improved designs, but by improving the methods by which designs are generated and evaluated. The analytical and empirical approaches employed in this volume currently exist in some state of tension in the research community. Nevertheless, most would grant that evidence from computational experiments, theoretical analysis, and real-world experience are all legitimately admissible in evaluation of trading agent strategies. Reconciling in a satisfactory way these sources of knowledge is one of the most fundamental challenges for the enterprise of agent design in coming years.

Bibliography

Kenneth J. Arrow and F. H. Hahn. *General Competitive Analysis*. Holden-Day, San Francisco, 1971. Cited on page(s) 65

Lawrence M. Ausubel and Peter Cramton. Demand reduction and inefficiency in multi-unit auctions. Working paper, University of Maryland, 2002. Cited on page(s) 45, 72

Thomas A. Bass. *The Predictors*. Henry Holt and Company, 1999. Cited on page(s) 6

Michael Benisch, Alberto Sardinha, James Andrews, and Norman Sadeh. CMieux: Adaptive strategies for competitive supply chain trading. In *Eighth International Conference on Electronic Commerce*, pages 47–58, Fredericton, NB, 2006. DOI: 10.1145/1150735.1150737 Cited on page(s) 65

Martin Bichler. *The Future of e-Markets: Multidimensional Market Mechanisms*. Cambridge University Press, 2001. Cited on page(s) 81

John Birge and Francois Louveaux. *Introduction to Stochastic Programming*. Springer, 1997. Cited on page(s) 63

Christopher M. Bishop. *Pattern Recognition and Machine Learning*. Springer, 2006. Cited on page(s) 6

Craig Boutilier, Moisés Goldszmidt, and Bikash Sabata. Sequential auctions for the allocation of resources with complementarities. In *Sixteenth International Joint Conference on Artificial Intelligence*, pages 527–534, Stockholm, 1999. Cited on page(s) 74

Justin Boyan and Amy Greenwald. Bid determination in simultaneous auctions: An agent architecture. In *Third ACM Conference on Electronic Commerce*, pages 210–212, Tampa, 2001. DOI: 10.1145/501158.501184 Cited on page(s) 58

Andrew Byde, Chris Preist, and Nicholas R. Jennings. Decision procedures for multiple auctions. In *First International Joint Conference on Autonomous Agents and Multi-Agent Systems*, Bologna, 2002. DOI: 10.1145/544862.544888 Cited on page(s) 74

Mark M. Bykowsky, Robert J. Cull, and John O. Ledyard. Mutually destructive bidding: The FCC auction design problem. *Journal of Regulatory Economics*, 17:205–228, 2000. DOI: 10.1023/A:1008122015102 Cited on page(s) 74

E. C. Capen, R. V. Clapp, and W. M. Campbell. Competitive bidding in high-risk situations. *Journal of Petroleum Technology*, 23:641–653, 1971. DOI: 10.2118/2993-PA Cited on page(s) 39

Ralph Cassady, Jr. *Auctions and Auctioneering*. University of California Press, 1967. Cited on page(s) 30, 39

Tanmoy Chakraborty and Michael Kearns. Market making and mean reversion. In *Twelfth ACM Conference on Electronic Commerce*, San Jose, 2011. DOI: 10.1145/1993574.1993622 Cited on page(s) 50

Shih-Fen Cheng, Evan Leung, Kevin M. Lochner, Kevin O'Malley, Daniel M. Reeves, and Michael P. Wellman. Walverine: A Walrasian trading agent. *Decision Support Systems*, 39:169–184, 2005. DOI: 10.1016/j.dss.2003.10.005 Cited on page(s) 65, 66

Dave Cliff. Minimal-intelligence agents for bargaining behaviours in market-based environments. Technical Report HP-97-91, HP Laboratories, 1997. Cited on page(s) 47

Dave Cliff. ZIP60: Further explorations in the evolutionary design of trader agents and online auction-market mechanisms. *IEEE Transactions on Evolutionary Computation*, 13:3–18, 2009. DOI: 10.1109/TEVC.2008.907594 Cited on page(s) 47

Adam Cohen. *The Perfect Store: Inside eBay*. Little, Brown, and Company, 2002. Cited on page(s) 9

John Collins, Wolfgang Ketter, and Norman Sadeh. Pushing the limits of rational agents: The trading agent competition for supply chain management. *AI Magazine*, 31(2):63–80, 2010. Cited on page(s) 76

Peter Cramton. Simultaneous ascending auctions. In Cramton et al. [2005]. Cited on page(s) 67

Peter Cramton, Yoav Shoham, and Richard Steinberg, editors. *Combinatorial Auctions*. MIT Press, 2005. Cited on page(s) 58, 81, 84

Peter C. Cramton. The FCC spectrum auctions: An early assessment. *Journal of Economics and Management Strategy*, 6:431–495, 1997. DOI: 10.1111/j.1430-9134.1997.00431.x Cited on page(s) 74

Rajarshi Das, James E. Hanson, Jeffrey O. Kephart, and Gerald Tesauro. Agent-human interactions in the continuous double auction. In *Seventeenth International Joint Conference on Artificial Intelligence*, pages 1169–1176, Seattle, WA, 2001. Cited on page(s) 49

Sanmay Das. A learning market-maker in the Glosten–Milgrom model. *Quantitative Finance*, 5: 169–180, 2005. DOI: 10.1080/14697680500148067 Cited on page(s) 50

Ian Domowitz. Automating the continuous double auction in practice: Automated trade execution systems in financial markets. In Friedman and Rust [1993], pages 27–60. Cited on page(s) 6

Charles Duhigg. Artificial intelligence applied heavily to picking stocks. *International Herald Tribune*, 23 November 2006. Cited on page(s) 6

Robert F. Easley and Rafael Tenorio. Jump bidding strategies in Internet auctions. *Management Science*, 50:1407–1419, 2004. DOI: 10.1287/mnsc.1040.0286 Cited on page(s) 29

Richard Engelbrecht-Wiggans and Robert J. Weber. An example of a multi-object auction game. *Management Science*, 25:1272–1277, 1979. DOI: 10.1287/mnsc.25.12.1272 Cited on page(s) 74

Michael J. Fishman. A theory of preemptive takeover bidding. *RAND Journal of Economics*, 19: 88–101, 1988. DOI: 10.2307/2555399 Cited on page(s) 29

Daniel Friedman. The double auction market institution: A survey. In Friedman and Rust [1993], pages 3–25. Cited on page(s) 41

Daniel Friedman and John Rust, editors. *The Double Auction Market: Institutions, Theories, and Evidence*. Addison-Wesley, 1993. Cited on page(s) 46, 84, 85, 89

Mark B. Garman. Market microstructure. *Journal of Financial Economics*, 3:257–275, 1976. DOI: 10.1016/0304-405X(76)90006-4 Cited on page(s) 6, 49

Evan Gatev, William N. Goetzmann, and K. Geert Rouwenhorst. Pairs trading: Performance of a relative-value arbitrage rule. *Review of Financial Studies*, 19:797–827, 2006. DOI: 10.1093/rfs/hhj020 Cited on page(s) 6

Enrico H. Gerding, Zinovi Rabinovich, Andrew Byde, Edith Elkind, and Nicholas R. Jennings. Approximating mixed Nash equilibria using smooth fictitious play in simultaneous auctions (short paper). In *Seventh International Conference on Autonomous Agents and Multi-Agent Systems*, pages 1577–1580, Estoril, Portugal, 2008. Cited on page(s) 74

Steven Gjerstad. The competitive market paradox. *Journal of Economic Dynamics and Control*, 31: 1753–1780, 2007. DOI: 10.1016/j.jedc.2006.07.001 Cited on page(s) 48

Steven Gjerstad and John Dickhaut. Price formation in double auctions. *Games and Economic Behavior*, 22:1–29, 1998. DOI: 10.1006/game.1997.0576 Cited on page(s) 48

Dhananjay K. Gode and Shyam Sunder. Allocative efficiency of markets with zero-intelligence traders: Market as a partial substitute for individual rationality. *Journal of Political Economy*, 101: 119–137, 1993. DOI: 10.1086/261868 Cited on page(s) 46

Amy Greenwald and Justin Boyan. Bidding algorithms for simultaneous auctions: A case study. In *Third ACM Conference on Electronic Commerce*, pages 115–124, Tampa, 2001. DOI: 10.1145/501158.501171 Cited on page(s) 74

Amy Greenwald and Justin Boyan. Bidding under uncertainty: Theory and experiments. In *Twentieth Conference on Uncertainty in Artificial Intelligence*, pages 209–216, Banff, 2004. Cited on page(s) 74

Amy Greenwald, Seong Jae Lee, and Victor Naroditskiy. RoxyBot-06: Stochastic prediction and optimization in TAC travel. *Journal of Artificial Intelligence Research*, 36:513–546, 2009. DOI: 10.1613/jair.2904 Cited on page(s) 64, 74

Amy Greenwald, Victor Naroditskiy, and Seong Jae Lee. Bidding heuristics for simultaneous auctions: Lessons from TAC travel. In Wolfgang Ketter, Han La Poutré, Norman Sadeh, Onn Shehory, and William Walsh, editors, *Agent-Mediated Electronic Commerce and Trading Agent Design and Analysis*, volume 44 of *Lecture Notes in Business Information Processing*, pages 131–146. Springer-Verlag, 2010. Cited on page(s) 74

Trevor Hastie, Robert Tibshirani, and Jerome Friedman. *Elements of Statistical Learning*. Springer-Verlag, 2001. Cited on page(s) 6

Lu Hong and Scott Page. Interpreted and generated signals. *Journal of Economic Theory*, 144: 2174–2196, 2009. DOI: 10.1016/j.jet.2009.01.006 Cited on page(s) 31, 39

John C. Hull. *Options, Futures, and Other Derivatives*. Prentice Hall, fourth edition, 2000. Cited on page(s) 4

Patrick R. Jordan and Michael P. Wellman. Generalization risk minimization in empirical game models. In *Eighth International Conference on Autonomous Agents and Multi-Agent Systems*, pages 553–560, Budapest, 2009. Cited on page(s) 51

Patrick R. Jordan, Michael P. Wellman, and Guha Balakrishnan. Strategy and mechanism lessons from the first ad auctions trading agent competition. In *Eleventh ACM Conference on Electronic Commerce*, Cambridge, MA, 2010. DOI: 10.1145/1807342.1807389 Cited on page(s) 7

Adam I. Juda and David C. Parkes. An options-based solution to the sequential auction problem. *Artificial Intelligence*, 173:876–899, 2009. DOI: 10.1016/j.artint.2009.01.002 Cited on page(s) 74

Michael Kearns and Luis Ortiz. The Penn-Lehman automated trading project. *IEEE Intelligent Systems*, 18(6):22–31, 2003. DOI: 10.1109/MIS.2003.1249166 Cited on page(s) 50

Jeffrey O. Kephart, James E. Hanson, and Jakka Sairamesh. Price and niche wars in a free-market economy of software agents. *Artificial Life*, 4:1–23, 1998. DOI: 10.1162/106454698568413 Cited on page(s) 68

Wolfgang Ketter, Han La Poutré, Norman Sadeh, and William Walsh, editors. *Agent-Mediated Electronic Commerce and Trading Agent Design and Analysis*, volume 44 of *Lecture Notes in Business Information Processing*. Springer-Verlag, 2010. Cited on page(s) 88

Christopher Kiekintveld, Jason Miller, Patrick R. Jordan, and Michael P. Wellman. Controlling a supply chain agent using value-based decomposition. In *Seventh ACM Conference on Electronic Commerce*, pages 208–217, Ann Arbor, MI, 2006. DOI: 10.1145/1134707.1134730 Cited on page(s) 76, 79

Christopher Kiekintveld, Jason Miller, Patrick R. Jordan, Lee F. Callender, and Michael P. Wellman. Forecasting market prices in a supply chain game. *Electronic Commerce Research and Applications*, 8:63–77, 2009. DOI: 10.1016/j.elerap.2008.11.005 Cited on page(s) 65, 78

Kendall Kim. *Electronic and Algorithmic Trading Technology*. Academic Press, 2007. Cited on page(s) 6

Brendan Kitts and Benjamin Leblanc. Optimal bidding on keyword auctions. *Electronic Markets*, 14:186–201, 2004. DOI: 10.1080/1019678042000245119 Cited on page(s) 7

Paul Klemperer. *Auctions: Theory and Practice*. Princeton University Press, 2004. Cited on page(s) 29, 39

Anton J. Kleywegt, Alexander Shapiro, and Tito Homem de Mello. The sample average approximation method for stochastic discrete optimization. *SIAM Journal on Optimization*, 12:479–502, 2002. DOI: 10.1137/S1052623499363220 Cited on page(s) 64

Sarit Kraus. *Strategic Negotiation in Multiagent Environments*. MIT Press, 2001. Cited on page(s) 6

Vijay Krishna. *Auction Theory*. Academic Press, second edition, 2010. Cited on page(s) 6, 24, 39

Vijay Krishna and Robert W. Rosenthal. Simultaneous auctions with synergies. *Games and Economic Behavior*, 17:1–31, 1996. DOI: 10.1006/game.1996.0092 Cited on page(s) 74

Sébastien Lahaie, David M. Pennock, Amin Saberi, and Rakesh V. Vohra. Sponsored search auctions. In Noam Nisan, Tim Roughgarden, Éva Tardos, and Vijay V. Vazirani, editors, *Algorithmic Game Theory*, pages 699–716. Cambridge University Press, 2007. Cited on page(s) 6, 82

Young Han Lee and Ulrike Malmendier. The bidder's curse. *American Economic Review*, 101: 749–787, 2011. DOI: 10.1257/aer.101.2.749 Cited on page(s) 13

Benny Lehmann, Daniel Lehmann, and Noam Nisan. Combinatorial auctions with decreasing marginal utilities. *Games and Economic Behavior*, 55:270–296, 2006. DOI: 10.1016/j.geb.2005.02.006 Cited on page(s) 69, 74

Kevin Leyton-Brown and Yoav Shoham. *Essentials of Game Theory: A Concise Multidisciplinary Introduction*. Morgan and Claypool, 2008. DOI: 10.2200/S00108ED1V01Y200802AIM003 Cited on page(s) xii

David Lucking-Reiley. Vickrey auctions in practice: From nineteenth-century philately to twenty-first century e-commerce. *Journal of Economic Perspectives*, 14(3):183–192, 2000. DOI: 10.1257/jep.14.3.183 Cited on page(s) 39

Jeffrey K. MacKie-Mason and Michael P. Wellman. Automated markets and trading agents. In Leigh Tesfatsion and Kenneth L. Judd, editors, *Handbook of Agent-Based Computational Economics*. Elsevier, 2006. Cited on page(s) 58

Jeffrey K. MacKie-Mason, Anna Osepayshvili, Daniel M. Reeves, and Michael P. Wellman. Price prediction strategies for market-based scheduling. In *Fourteenth International Conference on Automated Planning and Scheduling*, pages 244–252, Whistler, BC, 2004. Cited on page(s) 74

Andreu Mas-Colell, Michael D. Whinston, and Jerry R. Green. *Microeconomic Theory*. Oxford University Press, New York, 1995. Cited on page(s) 66

R. Preston McAfee and John McMillan. Auctions and bidding. *Journal of Economic Literature*, 25: 699–738, 1987. Cited on page(s) 39

Paul Milgrom. Auctions and bidding: A primer. *Journal of Economic Perspectives*, 3(3):3–22, 1989. Cited on page(s) 39

Paul Milgrom. Putting auction theory to work: The simultaneous ascending auction. *Journal of Political Economy*, 108:245–272, 2000. DOI: 10.1086/262118 Cited on page(s) 28, 68, 69, 74

Paul Milgrom. *Putting Auction Theory to Work*. Cambridge University Press, 2003. Cited on page(s) 39, 67

Paul R. Milgrom and Robert J. Weber. A theory of auctions and competitive bidding. *Econometrica*, 50:1089–1122, 1982. DOI: 10.2307/1911865 Cited on page(s) 39

Ross M. Miller. *Paving Wall Street: Experimental Economics and the Quest for the Perfect Market*. Wiley, 2002. Cited on page(s) 6, 43

Lonneke Mous, Valentin Robu, and Han La Poutré. Using priced options to solve the exposure problem in sequential auctions. In Ketter et al. [2010], pages 29–45. Cited on page(s) 74

Abhinay Muthoo. *Bargaining Theory with Applications*. Cambridge University Press, 1999. Cited on page(s) 2

Roger B. Myerson. Incentive compatibility and the bargaining problem. *Econometrica*, 47:61–73, 1979. DOI: 10.2307/1912346 Cited on page(s) 26

Yuriy Nevmyvaka, Yi Feng, and Michael Kearns. Reinforcement learning for optimized trade execution. In *Twenty-Third International Conference on Machine Learning*, pages 673–680, Pittsburgh, 2006. DOI: 10.1145/1143844.1143929 Cited on page(s) 53

Axel Ockenfels, David H. Reiley, and Abdolkarim Sadrieh. Online auctions. In Terrence Hendershott, editor, *Economics and Information Systems*. Elsevier, 2006. Cited on page(s) 16

Maureen O'Hara. *Market Microstructure Theory*. Blackwell, 1995. Cited on page(s) 6, 49, 50

David Pardoe and Peter Stone. An autonomous agent for supply chain management. In *Business Computing*, volume 3 of *Handbooks in Information Systems*, pages 141–172. Emerald Group, 2009. Cited on page(s) 65

David Pardoe and Peter Stone. The 2007 TAC SCM prediction challenge. In Ketter et al. [2010], pages 175–189. Cited on page(s) 65, 78

David Pardoe and Peter Stone. A particle filter for bid estimation in ad auctions with periodic ranking observations. In *EC-10 Workshop on Trading Agent Design and Analysis*, Cambridge, MA, 2010b. Cited on page(s) 65

David Pardoe, Doran Chakraborty, and Peter Stone. TacTex09: A champion bidding agent for ad auctions. In *Ninth International Conference on Autonomous Agents and Multi-Agent Systems*, pages 1273–1280, Toronto, 2010. Cited on page(s) 7

David C. Parkes and Jayant Kalagnanam. Models for iterative multiattribute procurement auctions. *Management Science*, 51:435–451, 2005. DOI: 10.1287/mnsc.1040.0340 Cited on page(s) 81

David W. Pearce, editor. *The Dictionary of Modern Economics*. MIT Press, 1983. Cited on page(s) 3

S. Phelps, M. Marcinkiewicz, S. Parsons, and P. McBurney. A novel method for automatic strategy acquisition in n-player non-zero-sum games. In *Fifth International Joint Conference on Autonomous Agents and Multi-Agent Systems*, pages 705–712, Hakodate, 2006.
DOI: 10.1145/1160633.1160760 Cited on page(s) 51

Chris Preist. Commodity trading using an agent-based iterated double auction. In *Third International Conference on Autonomous Agents*, pages 131–138, 1999. DOI: 10.1145/301136.301179 Cited on page(s) 47

Paul S. A. Reitsma, Peter Stone, János A. Csirik, and Michael L. Littman. Self-enforcing strategic demand reduction. In *Agent-Mediated Electronic Commerce IV*, volume 2531 of *Lecture Notes in Computer Science*, pages 113–129. Springer-Verlag, 2002. Cited on page(s) 74

Alvin E. Roth and Axel Ockenfels. Last-minute bidding and the rules for ending second-price auctions: Evidence from eBay and Amazon auctions on the Internet. *American Economic Review*, 92:1093–1103, 2002. DOI: 10.1257/00028280260344632 Cited on page(s) 13, 14

Stuart Russell and Peter Norvig. *Artificial Intelligence: A Modern Approach*. Prentice Hall, third edition, 2009. Cited on page(s) 1, 6, 47

John Rust, John H. Miller, and Richard Palmer. Characterizing effective trading strategies: Insights from a computerized double auction tournament. *Journal of Economic Dynamics and Control*, 18: 61–96, 1994. DOI: 10.1016/0165-1889(94)90069-8 Cited on page(s) 46

Felix Salmon and Jon Stokes. Algorithms take control of Wall Street. *Wired Magazine*, 27 December 2010. Cited on page(s) 6

Tuomas Sandholm. Issues in computational Vickrey auctions. *International Journal of Electronic Commerce*, 4:107–129, 2000. Cited on page(s) 26

Mark A. Satterthwaite and Steven R. Williams. The Bayesian theory of the k-double auction. In Friedman and Rust [1993], pages 99–123. Cited on page(s) 44

L. Julian Schvartzman and Michael P. Wellman. Market-based allocation with indivisible bids. *Production and Operations Management*, 16:495–509, 2007. DOI: 10.1111/j.1937-5956.2007.tb00275.x Cited on page(s) 43

L. Julian Schvartzman and Michael P. Wellman. Stronger CDA strategies through empirical game-theoretic analysis and reinforcement learning. In *Eighth International Conference on Autonomous Agents and Multi-Agent Systems*, pages 249–256, Budapest, 2009. Cited on page(s) 51

L. Julian Schvartzman and Michael P. Wellman. Learning improved entertainment trading strategies for the TAC travel game. In Esther David, Enrico Gerding, David Sarne, and Onn Shehory, editors, *Agent-Mediated Electronic Commerce: Designing Trading Strategies and Mechanisms for Electronic Markets*, volume 59 of *Lecture Notes in Business Information Processing*. Springer-Verlag, 2010. Cited on page(s) 53

Alexander Sherstov and Peter Stone. Three automated stock-trading agents: A comparative study. In *AAMAS-04 Workshop on Agent-Mediated Electronic Commerce*, New York, 2004. Cited on page(s) 50

Yoav Shoham and Kevin Leyton-Brown. *Multiagent Systems: Algorithmic, Game-Theoretic and Logical Foundations*. Cambridge University Press, 2009. Cited on page(s) 6, 28, 39

Ken Steiglitz. *Snipers, Shills, and Sharks: eBay and Human Behavior*. Princeton University Press, 2007. Cited on page(s) 16, 39

Peter Stone, Robert E. Schapire, Michael L. Littman, János A. Csirik, and David McAllester. Decision-theoretic bidding based on learned density models in simultaneous, interacting auctions. *Journal of Artificial Intelligence Research*, 19:209–242, 2003. DOI: 10.1613/jair.1200 Cited on page(s) 65

Katia Sycara and Tinglong Dai. Agent reasoning in negotiation. In *Handbook of Group Decision and Negotiation*, volume 4 of *Advances in Group Decision and Negotiation*, pages 437–451. Springer, 2010. DOI: 10.1007/978-90-481-9097-3_26 Cited on page(s) 6

Csaba Szepesvári. *Algorithms for Reinforcement Learning*. Morgan and Claypool, 2010. DOI: 10.2200/S00268ED1V01Y201005AIM009 Cited on page(s) 6, 52

Gerald Tesauro and Jonathan L. Bredin. Strategic sequential bidding in auctions using dynamic programming. In *First International Joint Conference on Autonomous Agents and Multi-Agent Systems*, pages 591–598, Bologna, 2002. DOI: 10.1145/544862.544885 Cited on page(s) 49

Gerald Tesauro and Rajarshi Das. High-performance bidding agents for the continuous double auction. In *Third ACM Conference on Electronic Commerce*, pages 206–209, Tampa, 2001. DOI: 10.1145/501158.501183 Cited on page(s) 47, 49, 51

Richard H. Thaler. The winner's curse. *Journal of Economic Perspectives*, 2(1):191–202, 1988. Cited on page(s) 39

William Vickrey. Counterspeculation, auctions, and competitive sealed tenders. *Journal of Finance*, 16:8–37, 1961. DOI: 10.2307/2977633 Cited on page(s) 2, 25, 39

Yevgeniy Vorobeychik, Michael P. Wellman, and Satinder Singh. Learning payoff functions in infinite games. *Machine Learning*, 67:145–168, 2007. DOI: 10.1007/s10994-007-0715-8 Cited on page(s) 51

Perukrishnen Vytelingum, Rajdeep K. Dash, Esther David, and Nicholas R. Jennings. A risk-based bidding strategy for continuous double auctions. In *Sixteenth European Conference on Artificial Intelligence*, pages 79–83, Valencia, Spain, 2004. Cited on page(s) 47

Perukrishnen Vytelingum, Dave Cliff, and Nicholas R. Jennings. Evolutionary stability of behavioural types in the continuous double auction. In *AAMAS-06 Joint Workshop on Trading Agent Design and Analysis and Agent Mediated Electronic Commerce*, Hakodate, 2006. Cited on page(s) 51

Perukrishnen Vytelingum, Dave Cliff, and Nicholas R. Jennings. Strategic bidding in continuous double auctions. *Artificial Intelligence*, 172:1700–1729, 2008. DOI: 10.1016/j.artint.2008.06.001 Cited on page(s) 47, 48

William E. Walsh, Rajarshi Das, Gerald Tesauro, and Jeffrey O. Kephart. Analyzing complex strategic interactions in multi-agent systems. In *AAAI-02 Workshop on Game-Theoretic and Decision-Theoretic Agents*, Edmonton, 2002. Cited on page(s) 50, 51

Robert J. Weber. Making more from less: Strategic demand reduction in the FCC spectrum auctions. *Journal of Economics and Management Strategy*, 6:529–548, 1997. DOI: 10.1162/105864097567183 Cited on page(s) 74

Michael P. Wellman and Peter R. Wurman. A trading agent competition for the research community. In *IJCAI-99 Workshop on Agent-Mediated Electronic Trading*, Stockholm, August 1999. Cited on page(s) 74

Michael P. Wellman, Daniel M. Reeves, Kevin M. Lochner, and Yevgeniy Vorobeychik. Price prediction in a trading agent competition. *Journal of Artificial Intelligence Research*, 21:19–36, 2004. DOI: 10.1145/779928.779966 Cited on page(s) 65

Michael P. Wellman, Daniel M. Reeves, Kevin M. Lochner, Shih-Fen Cheng, and Rahul Suri. Approximate strategic reasoning through hierarchical reduction of large symmetric games. In *Twentieth National Conference on Artificial Intelligence*, pages 502–508, Pittsburgh, 2005. Cited on page(s) 52

Michael P. Wellman, Amy Greenwald, and Peter Stone. *Autonomous Bidding Agents: Strategies and Lessons from the Trading Agent Competition*. MIT Press, 2007. Cited on page(s) 6, 60, 74

Michael P. Wellman, Anna Osepayshvili, Jeffrey K. MacKie-Mason, and Daniel M. Reeves. Bidding strategies for simultaneous ascending auctions. *B. E. Journal of Theoretical Economics (Topics)*, 8 (1), 2008. Cited on page(s) 69, 72, 74

Robert Wilson. On equilibria of bid-ask markets. In G. Feiwel, editor, *Arrow and the Ascent of Modern Economic Theory*, pages 375–414. MacMillan, 1987. Cited on page(s) 46

Ladislav Wintr. Some evidence on late bidding in eBay auctions. *Economic Inquiry*, 46:369–379, 2008. DOI: 10.1111/j.1465-7295.2007.00097.x Cited on page(s) 13, 14

Peter R. Wurman, William E. Walsh, and Michael P. Wellman. Flexible double auctions for electronic commerce: Theory and implementation. *Decision Support Systems*, 24:17–27, 1998. DOI: 10.1016/S0167-9236(98)00060-8 Cited on page(s) 42, 45

Peter R. Wurman, Michael P. Wellman, and William E. Walsh. A parametrization of the auction design space. *Games and Economic Behavior*, 35:304–338, 2001. DOI: 10.1006/game.2000.0828 Cited on page(s) 75

Dong Young Yoon and Michael P. Wellman. Self-confirming price prediction for bidding in simultaneous second-price sealed-bid auctions. In *IJCAI-11 Workshop on Trading Agent Design and Analysis*, Barcelona, 2011. Cited on page(s) 67

Author's Biography

MICHAEL P. WELLMAN

Michael P. Wellman is Professor of Computer Science & Engineering at the University of Michigan. He received a PhD from the Massachusetts Institute of Technology in 1988 for his work in qualitative probabilistic reasoning and decision-theoretic planning. From 1988–1992, Wellman conducted research in these areas at the USAF's Wright Laboratory. For the past 20 years, his research has focused on computational market mechanisms for distributed decision making and electronic commerce. As Chief Market Technologist for TradingDynamics, Inc. (now part of Ariba), he designed configurable auction technology for dynamic business-to-business commerce. Wellman previously served as Chair of the ACM Special Interest Group on Electronic Commerce (SIGecom), and as Executive Editor of the Journal of Artificial Intelligence Research. He is a Fellow of the Association for the Advancement of Artificial Intelligence and the Association for Computing Machinery.